Advancing Maths for AQA
MECHANICS 1

Ted Graham, Aidan Burrows and Joan Corbett

Series editors
Ted Graham   **Sam Boardman**   **David Pearson**
Roger Willi...   College

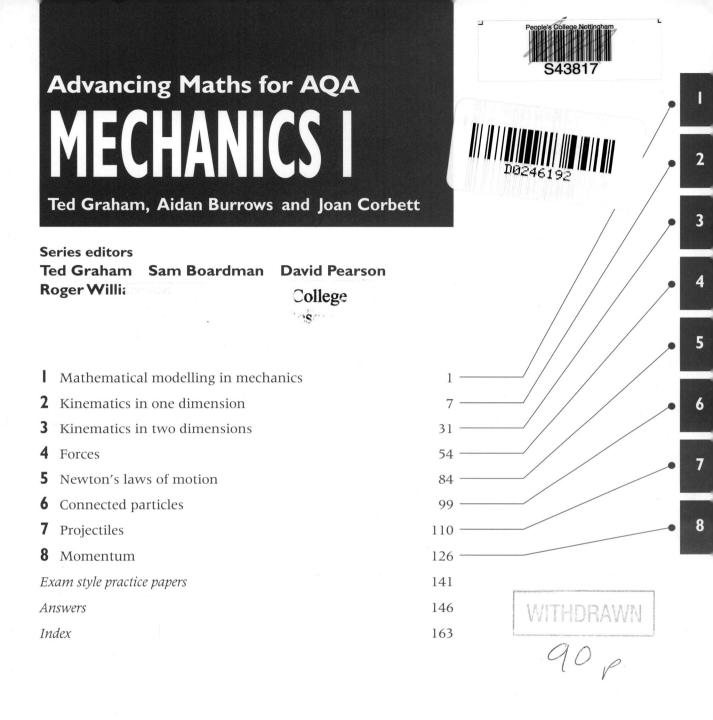

www.heinemann.co.uk
✓ Free online support
✓ Useful weblinks
✓ 24 hour online ordering

01865 888058

Heinemann
Inspiring generations

Heinemann Educational Publishers
Halley Court, Jordan Hill, Oxford OX2 8EJ
Part of Harcourt Education

Heinemann is the registered trademark of
Harcourt Education Limited

Text © Ted Graham, Aidan Burrows and Joan Corbett 2000, 2004
Complete work © Heinemann Educational Publishers 2004

First published in 2004

08 07 06 05 04
10 9 8 7 6 5 4 3 2 1

British Library Cataloguing in Publication Data is available
from the British Library on request.

ISBN 0 435 51336 2

Edited by Alex Sharpe, Standard Eight Limited
Typeset and illustrated by Tech-Set Limited, Gateshead, Tyne & Wear
Original illustrations © Harcourt Education Limited, 2004
Cover design by Miller, Craig and Cocking Ltd
Printed in Great Britain by Scotprint

Acknowledgements
The publishers and authors acknowledge the work of the writers, Ray Atkin,
John Berry, Derek Collins, Tim Cross, Ted Graham, Phil Rawlins, Tom Roper,
Rob Summerson, Nigel Price, Frank Chorlton and Andy Martin of the *AEB
Mathematics for AQA A-Level* Series, from which some exercises and examples
have been taken.

The publishers' and authors' thanks are due to the AQA for permission to
reproduce questions from past examination papers.

The answers have been provided by the authors and are not the responsibility
of the examining board.

Every effort has been made to contact copyright holders of material reproduced
in this book. Any omissions will be rectified in subsequent printings if notice is
given to the publishers.

About this book

This book is one in a series of textbooks designed to provide you with exceptional preparation for AQA's new Advanced GCE Specification. The series authors are all senior members of the examining team and have prepared the textbooks specifically to support you in studying this course.

Finding your way around

The following are there to help you find your way around when you are studying and revising:

- **edge marks** (shown on the front page) – these help you to get to the right chapter quickly;
- **contents list** – this identifies the individual sections dealing with key syllabus concepts so that you can go straight to the areas that you are looking for;
- **index** – a number in bold type indicates where to find the main entry for that topic.

Key points

Key points are not only summarised at the end of each chapter but are also boxed and highlighted within the text like this:

> For every action, there is an equal but opposite reaction

Exercises and exam questions

Worked examples and carefully graded questions familiarise you with the specification and bring you up to exam standard. Each book contains:

- Worked examples and Worked exam questions to show you how to tackle typical questions;
- Graded exercises, gradually increasing in difficulty up to exam-level questions, which are marked by an [A];
- Test-yourself sections for each chapter so that you can check your understanding of the key aspects of that chapter and identify any sections that you should review;
- Answers to the questions are included at the end of the book.

Mathematical modelling in mechanics

Learning objectives

After studying this chapter you should be able to:
- apply the mathematical modelling cycle to simple problems
- understand some of the assumptions used when modelling in a mechanics context
- understand some of the differences between a particle model and a rigid body model
- use some of the terminology found in mechanics.

1.1 Introducing modelling

Mathematical modelling describes the process of obtaining a solution to a real world problem. Many real problems can be very complex and so the idea of creating a mathematical model is to simplify the real situation, so that it can be described using equations or graphs. These equations or graphs are referred to as a mathematical model. These mathematical models can provide solutions to the original problem. It is often necessary to interpret these answers in the context of the original problem and to check that the answers that you have obtained are reasonable.

Examples of problems where modelling could be used:

- to determine the maximum speed of a car round a bend,
- to help define the design requirements of a sports stadium,
- to evaluate new design options for a mountain bike,
- to work out how to send a space station into orbit.

> This whole process is often referred to as the mathematical modelling cycle because it may be necessary to repeat the process, creating better models of reality until a satisfactory solution is obtained.

The key stages of the mathematical modelling cycle are shown in the diagram on the next page.

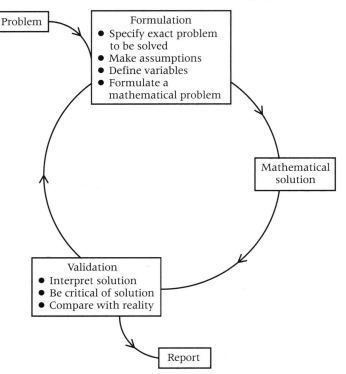

The mathematical modelling cycle

Each stage of this cycle is now considered in more detail.

1.2 Formulation

This stage of the modelling cycle consists of three distinct activities:

- specifying the exact problem to be solved,
- making assumptions,
- defining the variables you need to include,
- formulating a mathematical problem.

Specify problem

This is the starting point of the modelling cycle. Very often these problems will be very open-ended, unlike the types of problems that you will find in textbook exercises. Your first task is to make sure that you understand the problem and have decided what you need to find to solve the problem.

Make assumptions

The next part of the formulation stage is to make assumptions to simplify the problem. Some of the assumptions that you will often make in mechanics are now considered in the context of a body that remains at rest or moves in some way.

Assume that the body under consideration is a **particle**. This means that you assume that the body has no size, but does have a mass. So any rotation of the body is ignored, any forces will all act in the one place and the position is precisely defined. It may be quite reasonable to model a car as a particle if it is being driven a large distance or to model a ball as a particle.

A more sophisticated model is that of a **rigid body**. Here the body is assumed to have size, but to be rigid, so that it does not compress or change shape. The simplest rigid body is a rod. With a rigid body the forces will probably not all act at the same place and this must be taken into account. If you want to model the motion of a rolling ball it will be important to consider its size and shape, so a rigid sphere may be a suitable body.

> A particle has no size but does have a mass.
>
> A rigid body has size but does not change shape when forces are applied to it.

There are some factors, such as air resistance, lift on a golf ball, friction, resistance in a pulley, that may make a problem hard to describe and solve mathematically. We often ignore these factors in order to develop a simple mathematical model, but we may have to incorporate them at a later stage if our results do not give good predictions. Further examples of things that we might ignore are the mass of a string, friction, and the fact that a string might be elastic and stretch.

Any assumptions made should be clearly recorded so that they can be reviewed later in the process.

> Some keywords that you will find have particular meanings in a modelling context are listed below:
>
> - smooth   no friction,
> - rough   friction present,
> - light   has no mass,
> - inelastic   does not stretch,
> - inextensible   does not stretch.

Defining variables

You may also need to introduce algebraic variables to represent physical quantities that are important in the context of the problem. For example:

    m the mass of the body

    u the initial velocity of the body,
       etc.

Finally in the formulation stage there are a number of standard mathematical models to describe physical phenomena, for example to calculate the friction present, the size of the gravitational attraction, the force exerted by a stretched spring. You will learn about these models as you work through the mechanics modules and be able to include them in any modelling that you do.

Formulate a mathematical problem

In this phase of the modelling cycle you turn the open-ended problem that you started with into a focused mathematical problem that you can solve using mathematical techniques.

You may also need to collect some data that is relevant to the problem that you have to solve.

Mathematical solution

This is where you obtain an actual solution, by carrying out calculations, solving equations and using other mathematical techniques. As you learn more pure mathematics you will be able to deal with more sophisticated mathematical models. When you have obtained a solution you move on to the interpretation and validation phase.

Validation

This is where you interpret the mathematical answers in the context of the original problem. This is where you will probably reach a conclusion of some kind. For example you may state that it is not safe for a lorry to drive over a bridge, that a car was breaking a speed limit or that it is impossible to hit a tennis ball over the net from a certain position.

It is also important to validate your answer, to check that it is reasonable. This may require you to make some observations of reality or to carry out an experiment to validate your predictions or it may simply be that the model you have created produced totally unrealistic solutions. For example if you calculate that a cyclist rolling down a hill will reach a speed of 70 mph after travelling 100 m, by ignoring air resistance, then it is clear that you need to go back to your original assumptions and try to include air resistance.

If you do need to go back to your original assumptions and reformulate your model, then you are embarking on a second cycle, which should lead to a more realistic solution. Finally when you obtain a satisfactory solution, you will need to report on your conclusion, describing how you reached this conclusion.

Worked example 1.1

How far apart should speed bumps be placed so that traffic does not reach a speed greater than 30 mph?

Solution

Assume that:

- A car is a particle.
- The road conditions are good.
- The road is horizontal.
- The speed of the car is zero when it crosses the bumps.
- All cars slow down according to the table of stopping distances in the Highway Code.
- All cars speed up at the same rate as they slow down.

Gathering data from the Highway Code gives a figure of 23 m as the stopping distance at 30 mph.

Based on our assumptions we formulate the mathematical model:

Distance between speed bumps = 2 × stopping distance

Using this model we can calculate the required distance as 46 m (the distance to speed up to 30 mph and to slow down from 30 mph).

A person solving this problem in real life may be able to set up some experimental bumps and observe the speeds of cars between them.

There are some ways in which this model could be revised or reformulated, for example:

- Modelling the car as a rigid body that has length.
- Looking for alternative models to describe how a car gains speed and slows down.
- The car crosses the bumps at a low speed, for example 5 mph.

EXERCISE 1A

1 Gather some data on the lengths of cars and revise the solution to the speed bumps problem to take account of this factor.

2 Describe how changing the speed at which crossing the bumps would change the solution to the speed bumps problem.

3 Make a list of the assumptions that you might make to create a simple model of the motion of a javelin.

4 If you were to model the motion of the pendulum in a grandfather clock:
   (a) make a list of the assumptions that you might make;
   (b) make a list of the variables that you might include in your model.

5 A string that passes over a pulley connects two objects of equal masses. The initial positions of the objects are shown in the diagram. Discuss the different predictions that you would obtain if you model:

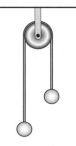

  (a) the pulley as smooth;

  (b) the pulley as smooth and the string as light.

6 A student models a parachutist as a particle that does not experience air resistance. Suggest what predictions he might obtain.

Key point summary

1 The mathematical modelling cycle consists of: *p1*

  ● formulating the problem;
  ● obtaining a mathematical solution;
  ● validating and interpreting the solution;
  ● preparing a report.

2 Modelling requires assumptions to be made.   *p2*

3 Particle and rigid body models are different as a particle has no size but has a mass, whereas a rigid body has size but does not change shape.   *p3*

4 Terminology used in mechanics: *p3*

  ● Smooth   no friction,
  ● Rough   friction present,
  ● Light   has no mass,
  ● Inelastic   does not stretch,
  ● Inextensible   does not stretch.

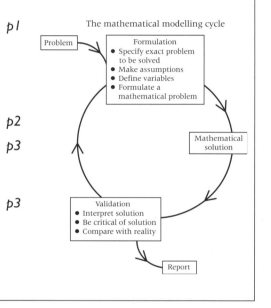

The mathematical modelling cycle

Problem

Formulation
● Specify exact problem to be solved
● Make assumptions
● Define variables
● Formulate a mathematical problem

Mathematical solution

Validation
● Interpret solution
● Be critical of solution
● Compare with reality

Report

CHAPTER 2
Kinematics in one dimension

Learning objectives

After studying this chapter you should be able to:
- define displacement, velocity and acceleration
- understand and interpret displacement–time graphs and velocity–time graphs
- solve problems involving motion under constant acceleration using standard formulae
- solve problems involving vertical motion under gravity.

2.1 Introducing kinematics

Kinematics is the study of motion and in this chapter we will study objects which move in one dimension only, this means that they move in a straight line. Examples are:

- cars, buses, or bikes, etc. on a straight road,
- objects dropped from the top of a cliff or tower so that they fall vertically,
- an athlete running in a 100 m race.

We will not, at this stage, consider what makes the objects move!

We shall also model most of the objects as particles so that we can ignore their size and shape and concentrate on their movement.

2.2 Displacement, velocity and acceleration

First, we have to define some words and symbols that we will use. In everyday language we use the terms **speed** and **distance** when talking about motion. **Distance** is how far we travel (miles, metres, etc.) and **speed** is how fast we go (miles per hour, metres per second, etc.). In the study of kinematics we need to be more precise and so we introduce **displacement** and **velocity**.

Displacement is based on the distance from a specific origin or reference point, but it also takes account of the direction in which the particle has moved.

The displacement may be 5 km north from a reference point or origin; or it may be given using positive or negative values relative to an origin, as shown in the diagram.

We decide that, relative to the point O, displacements to the right are positive and those to the left negative. Thus the displacement of P is $+3$ cm, and of Q -2 cm.

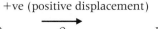
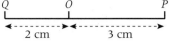

In mechanics the symbol used for displacement is s.

Velocity is defined similarly using the speed of an object together with the direction of the motion.

The velocity could be 6 mph going south-west from a reference point, or positive and negative values can be used.

In the diagram two particles are moving vertically. A is going upwards at 4 m s^{-1} and B is falling at 5 m s^{-1}. If we choose upwards as the positive direction, then the velocity of A is $+4$ m s^{-1} and B is -5 m s^{-1}.

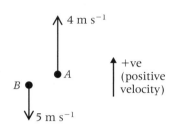

The symbol used for velocity is v.

We often speak of **average speed** meaning the **constant speed** we could have travelled at in order to cover a journey in the same time. For example, if you travel a journey of 100 miles in 2 hours, your average speed is 50 mph; it is very unlikely that you could have driven at a constant 50 mph!

$$\text{Average speed} = \frac{\text{Distance travelled}}{\text{Time taken}}$$

$$\text{Average velocity} = \frac{\text{Change in displacement}}{\text{Time taken}}$$

For example, if a particle moves 5 m from the origin to a point and then back to the origin in a time of 4 s:

$$\text{Average speed} = \frac{10}{4} = 2.5 \text{ m s}^{-1}$$

but

$$\text{Average velocity} = \frac{0}{4} = 0 \text{ m s}^{-1}.$$

For any motion that starts and finishes in the same place, the average velocity will be zero.

(Quantities that have size (magnitude) but no specific direction, such as speed and distance, are called **scalars**, but those with direction, like velocity and displacement, are called **vectors**. Vectors are studied in more detail in Chapter 3.)

Displacement–time graphs

As a body moves the displacement changes so that s is a function of time, t. A graph plotting s against t is called a displacement–time graph. As an example the graph shows the displacement–time graph for the motion of a ball thrown in the air and falling back to the floor.

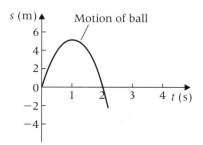

The motion of the ball can be described by looking at the graph. The ball starts at the origin and begins to move in the positive *s* direction (upwards) to a maximum height of 5 m above the point of release. It then falls to the floor which is 2 m below the point of release. It takes just over 2 seconds to hit the floor.

While the ball is moving upwards its velocity is positive. While it is moving downwards its velocity is negative. Note that the velocity is given by the gradient of the displacement–time graph.

In another example of a displacement–time graph, the table shows the displacement (*s*) of a boy, who is running on a straight track, measured at 2-second intervals.

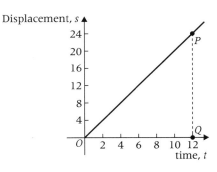

Time, *t*	0	2	4	6	8	10	12
Displacement, *s*	0	4	8	12	16	20	24

These values have been plotted on the graph. You notice that this graph is a straight line which tells us that the boy is running with a constant velocity.

His velocity is calculated from the gradient of the graph

$$\text{gradient} = \frac{PQ}{OQ} = \frac{24}{12} = 2$$

So the boy runs at 2 m s^{-1}.

Velocity–time graphs

If we know the velocity, *v*, at time *t*, then we can draw a graph of velocity against time.

This velocity–time graph is for the motion of the ball whose displacement–time graph we saw above. The ball is thrown up into the air with a velocity of 10 m s^{-1} and this decreases to zero as the ball reaches the highest point, in about 1 second. The direction of motion is then reversed as the ball falls back to the floor, so for this part of the motion the velocity is negative. The ball hits the floor with a velocity of about -12 m s^{-1}.

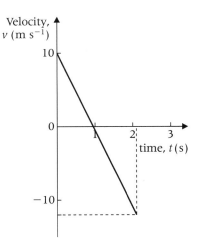

Here is another example of a velocity–time graph:

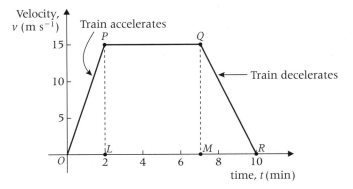

The velocity–time graph here is for the motion of a train which starts from rest at a station. Its velocity increases uniformly to 15 m s^{-1} in 2 minutes. Then it travels at that speed for 5 minutes before slowing down uniformly over a 3-minute period to stop at the next station.

Acceleration

Another familiar word used to describe the motion of cars, trains, bikes, and so on, in everyday language is acceleration. In the last two graphs above the velocity changes and the rate of change is called the acceleration.

Acceleration is defined as the rate at which the velocity is changing. Its units are 'metres per second per second' or m s^{-2}. So an acceleration of 5 m s^{-2} means that the velocity is increasing by 5 m s^{-1} every second.

> We can calculate the acceleration from the gradient of a velocity–time graph. Look at the v–t graph of the train above.

During the first part of the motion:

$$\text{acceleration} = \text{gradient of } OP$$
$$= \frac{15}{120}$$
$$= \frac{1}{8} \text{ or } 0.125 \text{ m s}^{-2}$$

Note that the time has been converted from minutes to seconds.

During the middle section the acceleration is zero.

In the third section of the train's journey it is slowing down; this is 'decelerating' or 'retarding' and the acceleration will have a negative value.

$$\text{acceleration} = \text{gradient of } QR$$
$$= \frac{-15}{180}$$
$$= -\frac{1}{12} \text{ or } -0.0833 \text{ m s}^{-2} \text{ (to three significant figures)}$$

Note that the symbol used for acceleration is a.

Displacement and the velocity–time graph

If a lorry moves at a constant speed of 15 m s^{-1} for 10 seconds, how far does it travel? The distance that the lorry travels is simply given by

$$15 \times 10 = 150 \text{ m}$$

If we look at the velocity–time graph for the lorry, we can see that 150 is the area of the rectangle *OABC*. We refer to this as the 'the area under the graph'.

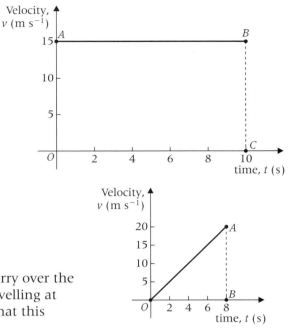

This method of calculating distance travelled from a *v–t* graph can be extended to cases where the velocity is not constant.

Later the lorry accelerates from rest to 20 m s^{-1} in 8 seconds. The diagram shows its *v–t* graph.

The area under this graph is the area of the triangle *OAB*, which can be calculated as

$$\text{Area} = \frac{20 \times 8}{2}$$
$$= 80$$

We can compare this with the average speed of the lorry over the 8-second period which would be 10 m s^{-1}. A lorry travelling at 10 m s^{-1} for 8 seconds would cover 80 metres. Note that this distance is the same as our area.

(This example is not intended to be a 'proof', but merely an indication of how the method works.)

> The area under a velocity–time graph represents the distance travelled.

Care must be taken with problems that include both positive and negative velocities. Consider the graph shown here. The shaded area marked A_1 represents a distance travelled in the positive direction, while the area marked A_2 represents a distance travelled in the negative direction. Comparing the sizes of the shaded areas indicates that a greater distance has been travelled in the positive direction than in the negative direction, so that the final displacement will be positive.

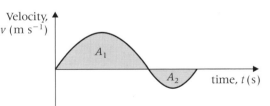

Worked example 2.1

Alongside a railway track there are marker posts, spaced at one kilometre intervals. A train, travelling at constant velocity, is timed to take 2 minutes to travel from one post to the next. After passing the second post, the train slows down uniformly to stop in 0.4 km. Find:

(a) the constant velocity of the train,

(b) the time it takes to stop,

(c) the acceleration of the train.

Solution

(There is a mixture of units in this question so we will use *seconds* for the time and *metres* for distance.)

First we make a sketch of the v–t graph, as shown. Note that BC is a straight line as the train slows down uniformly.

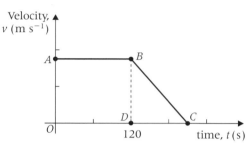

(a) The train travels 1000 m in 120 seconds, with a constant velocity, v.

Since the area under graph is equal to the distance travelled.

$$1000 = 120v$$

$$v = \frac{1000}{120} = \frac{25}{3} = 8.33 \text{ m s}^{-1} \text{ (to three significant figures)}.$$

(b) During the retardation, the train travels 400 m so the area of the triangle BDC must be 400 units, hence

$$\frac{BD \times DC}{2} = 400$$

$$DC = \frac{800}{BD} = 800 \times \frac{3}{25} = 96$$

Thus the train slows down for 96 seconds.

(c) The acceleration of the train is the gradient of BC.

$$a = \frac{-8.33}{96}$$

$$= -0.0868 \text{ m s}^{-2} \text{ (to three significant figures)}$$

Worked example 2.2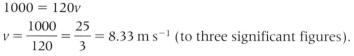

A cyclist rides along a straight road from X to Y. He starts from rest at X and accelerates uniformly to reach a speed of 10 m s^{-1} in 8 seconds. He travels at this speed for 20 seconds and then decelerates uniformly to stop at Y. If the whole journey takes 40 seconds, sketch a velocity–time graph for the journey.

Use the graph to find:

(a) the acceleration during the first 8 seconds,

(b) the acceleration on the final stage,

(c) the total distance travelled.

2

Solution

The *v–t* graph is shown below.

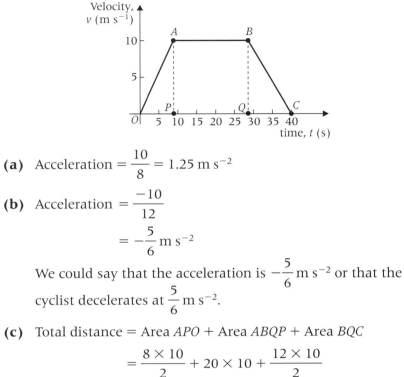

(a) Acceleration $= \dfrac{10}{8} = 1.25 \text{ m s}^{-2}$

(b) Acceleration $= \dfrac{-10}{12}$

$= -\dfrac{5}{6} \text{ m s}^{-2}$

We could say that the acceleration is $-\dfrac{5}{6} \text{ m s}^{-2}$ or that the cyclist decelerates at $\dfrac{5}{6} \text{ m s}^{-2}$.

(c) Total distance = Area *APO* + Area *ABQP* + Area *BQC*

$= \dfrac{8 \times 10}{2} + 20 \times 10 + \dfrac{12 \times 10}{2}$

$= 40 + 200 + 60 = 300 \text{ m}.$

EXERCISE 2A

1 The graph shows how the velocity of a car changes during a short journey along a straight road. Find the distance travelled by the car and the acceleration on each stage of its journey.

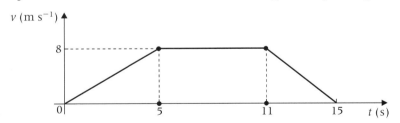

2 This graph shows how the velocity of a cyclist changes over a short period of time as she travels along a straight road. Find the total distance travelled by the cyclist.

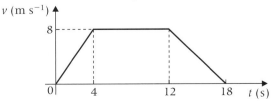

3 Discuss the motion represented by each of the displacement–time graphs shown here.

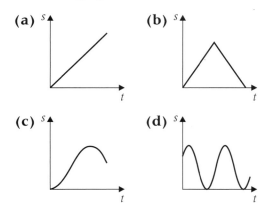

(a)   **(b)**

(c)   **(d)**

4 Sketch displacement–time and velocity–time graphs for the following.

(a) A car that starts from rest and increases its velocity steadily to 10 m s^{-1} in 5 seconds. The car holds this velocity for another 10 seconds and then slows steadily to rest in a further 10 seconds. Assume that the car travels along a straight line.

(b) A ball that is dropped on to a horizontal floor from a height of 3 m. The ball bounces several times before coming to rest.

(c) A person who jumps out of a balloon and falls until the parachute opens. The person then glides steadily to the ground. All the motion takes place along a vertical line.

 5

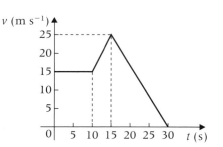

This velocity–time graph illustrates the motion of an object travelling along a straight line. Calculate the acceleration for each of the following intervals:

(a) $0 < t < 10$,

(b) $10 < t < 15$,

(c) $15 < t < 30$.

(d) Calculate the distance travelled by the object in the 30 seconds.

2

6 A car moves along a straight road. It accelerates at $2 \, \mathrm{m \, s^{-2}}$ from rest until it reaches a speed of $16 \, \mathrm{m \, s^{-1}}$. It then travels at this speed for 30 seconds, before slowing down and stopping in a further 5 seconds.

(a) Sketch a velocity–time graph for the car.

(b) Find the total distance travelled by the car.

7

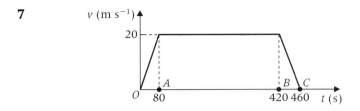

The diagram shows the velocity–time graph for a train which travels from rest in one station along a straight track to rest at the next station. For each of the time intervals *OA*, *AB* and *BC*, state the value of the train's acceleration.

Calculate the distance between the stations. [A]

8 The velocity–time graph shows the motion of a car and a bicycle as they travel along a straight horizontal road. When $t = 0$, the car and bicycle pass a traffic light on the road. At the traffic light, the bicycle is travelling at a constant velocity of $5 \, \mathrm{m \, s^{-1}}$, but the car is travelling at $3 \, \mathrm{m \, s^{-1}}$ and accelerating.

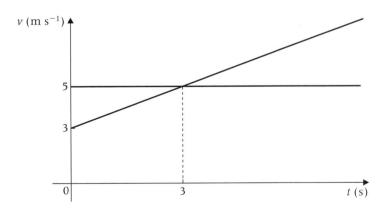

(a) (i) Explain how the graph indicates that the acceleration of the car is constant.

(ii) Find the acceleration of the car.

(b) When $t = T$, the car has travelled twice as far from the traffic light as the bicycle. Find the value of *T*. [A]

9 The velocity–time graph is for a train moving on a set of tracks over a 30-second period.

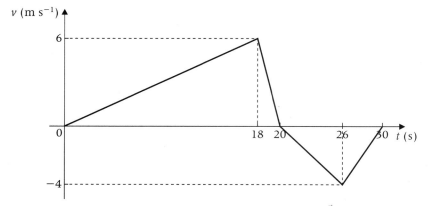

(a) Find the total distance travelled by the train.

(b) Calculate the average speed of the train.

(c) Calculate the average velocity of the train.

10 The displacement–time graph is for a lift.

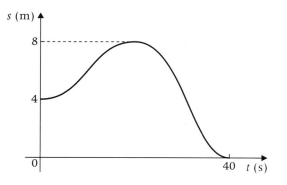

(a) Calculate the average speed of the lift during the 40-second period.

(b) Calculate the average velocity of the lift during the 40-second period.

11 A student attempts to model the motion of a bungee jumper. He draws the velocity–time graph shown:

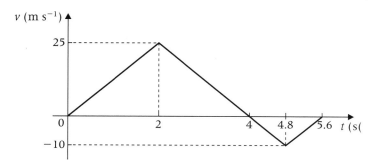

(a) State the two non-zero times at which the velocity of the bunjee jumper is zero.

(b) Find the distance that the bunjee jumper falls during the first 4 seconds.

(c) Find the total distance travelled by the bunjee jumper during the 5.6 seconds illustrated on the graph.   [A]

12 Two sprinters compete in a 100 m race, crossing the finishing line together after 12 seconds. The two models, A and B, as described below, are models for the motions of the two sprinters.

Model A.   The sprinter accelerates from rest at a constant rate for 4 seconds and then travels at a constant speed for the rest of the race.

Model B.   The sprinter accelerates from rest at a constant rate until reaching a speed of $9\,\text{m s}^{-1}$ and then travels at this speed for the rest of the race.

(a) Sketch velocity–time graphs for each model.

(b) For model A, find the maximum speed and the initial acceleration of the sprinter.

(c) For model B, find the time taken to reach the maximum speed and the initial acceleration of the sprinter.   [A]

13 The velocity–time graph shows the motion of a particle P moving with constant acceleration.

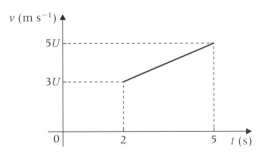

At times $t = 2$ and $t = 5$, P has velocities $3U$ and $5U$, respectively.

(a) (i) Find, in terms of U, the acceleration of P.

   (ii) Find, in terms of U, the distance travelled by P between the times $t = 2$ and $t = 5$.

(b) When $t = 5$, the motion of P changes and subsequently P moves with constant retardation. The particle P comes to rest after travelling a **further** 20 metres in the next 4 seconds. Find the value of U.   [A]

2.3 Motion under constant acceleration

There are several simple formulae which can be used to solve problems that involve motion under **constant** (or **uniform**) acceleration.

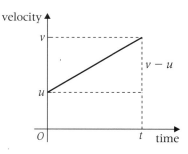

The diagram shows a velocity–time graph for the motion of an object with initial velocity u and final velocity v after t seconds has elapsed.

The gradient of the line is equal to the acceleration and is given by

$$\frac{v - u}{t}$$

Hence

$$a = \frac{v - u}{t}$$

which can be rewritten as

$$v = u + at.$$

The area under the velocity–time graph is equal to the displacement of the object. Using the rule for the area of a trapezium gives

$$s = \frac{1}{2}(u + v)t$$

Worked example 2.3

A motorbike accelerates at a constant rate of 3 m s^{-2}. Calculate:

(a) the time taken to accelerate from 18 km h^{-1} to 45 km h^{-1},

(b) the distance, in metres, covered in this time.

Solution

We can use the equation $v = u + at$ to find the time and then the equation $s = \frac{1}{2}(u + v)t$ to find the distance travelled. But first the units for speed must be converted to m s^{-1}.

$$18 \text{ km h}^{-1} = \frac{18 \times 1000}{3600} = 5 \text{ m s}^{-1}$$

And similarly

$$45 \text{ km h}^{-1} = \frac{45 \times 1000}{3600} = 12.5 \text{ m s}^{-1}.$$

(a) Using $v = u + at$, with $u = 5$, $v = 12.5$ and $a = 3$ gives

$$v = u + at$$
$$12.5 = 5 + 3t$$
$$t = \frac{7.5}{3} = 2.5 \text{ s}$$

(b) Using $s = \frac{1}{2}(u+v)t$ with $u = 5$, $v = 12.5$ and $t = 2.5$ gives

$$s = \frac{1}{2}(u+v)t$$

$$= \frac{1}{2}(12.5 + 5) \times 2.5$$

$$= 21.875 \text{ m}$$

Three more useful formulae

We can write the equation $v = u + at$ in the form

$$t = \frac{v-u}{a}$$

and substitute it into $s = \frac{1}{2}(u+v)t$, then

$$s = \frac{1}{2}(u+v)\left(\frac{v-u}{a}\right)$$

$$= \frac{v^2 - u^2}{2a}$$

which we can rearrange to give

$$v^2 = u^2 + 2as$$

Similarly, using $v = u + at$ to substitute for v in the equation

$$s = \frac{1}{2}(u+v)t$$

gives

$$s = \frac{(u + u + at)t}{2}$$

or

$$s = ut + \tfrac{1}{2}at^2$$

A final formula can be derived by eliminating u.

Using $u = v - at$ with $s = \frac{1}{2}(u+v)t$ gives

$$s = \frac{1}{2}(v + v - at)t$$

$$= vt - \frac{1}{2}at^2$$

This gives five different constant acceleration formulae. The appropriate one can be selected to fit the data available in the problem that you have to solve.

> When using these formulae it is important to remember that they only apply to cases where the acceleration is constant or can be assumed to be constant.

Worked example 2.4

A car accelerates uniformly from a velocity of 16 m s^{-1} to a velocity of 40 m s^{-1} while it travels a distance of 500 m along a straight road. Find the acceleration of the car.

Solution

Using the equation $v^2 = u^2 + 2as$, with $u = 16$, $v = 40$ and $s = 500$ gives

$$40^2 = 16^2 + 2 \times a \times 500$$

$$a = \frac{1600 - 256}{1000} = 1.344 \text{ m s}^{-2}$$

Worked example 2.5

A car decelerates from a velocity of 36 m s^{-1} while travelling along a straight road. The magnitude of the deceleration is 3 m s^{-2}. Calculate the time required to travel a distance of 162 m.

Solution

When an object is slowing down (decelerating) we can use the constant acceleration equations, but with a negative value for a. In this case $a = -3$.

We require the time, t seconds, to travel 162 m.

Using the constant acceleration equation $s = ut + \frac{1}{2}at^2$, with $s = 162$, $u = 36$ and $a = -3$ gives

$$162 = 36t + \tfrac{1}{2}(-3)t^2$$

Rearranging this gives this quadratic equation:

$$1.5t^2 - 36t + 162 = 0$$

Dividing by 1.5 gives

$$t^2 - 24t + 108 = 0$$

And factorising gives

$$(t - 6)(t - 18) = 0$$

so that

$$t = 6 \text{ or } t = 18$$

The first answer, of 6 seconds, is the required one.

The second answer would give the time that the displacement was 162 m for the second time. In this case the car would be moving in the opposite direction.

Worked example 2.6

A cyclist is initially travelling at $10 \, \text{m s}^{-1}$, when she applies her brakes. Assume that her acceleration remains constant at $-0.8 \, \text{m s}^{-2}$ until she stops and that she travels along a straight line. Find the distance that she travels before stopping and the time that it takes her to stop.

Solution

To find the distance that she travels use the formula $v^2 = u^2 + 2as$, with $v = 0$, $u = 10$ and $a = -0.8$, which gives

$$v^2 = u^2 + 2as$$
$$0^2 = 10^2 + 2 \times (-0.8)s$$
$$0 = 100 - 1.6s$$
$$s = \frac{100}{1.6} = 62.5 \, \text{m}$$

The time taken to stop can be found using the formula $v = u + at$ with $a = -0.8$, $u = 10$ and $v = 0$, to give

$$v = u + at$$
$$0 = 10 - 0.8t$$
$$0.8t = 10$$
$$t = \frac{10}{0.8} = 12.5 \, \text{s}$$

EXERCISE 2B

1 A car accelerates at $2 \, \text{m s}^{-2}$ from rest for 10 seconds along a straight road.

   (a) Find the distance travelled by the car and the speed it reaches.

   (b) After the 10 seconds its acceleration changes to $0.5 \, \text{m s}^{-2}$ and then remains constant for a further 5 seconds. Find the speed of the car and the total distance that it has travelled at the end of the 15 seconds.

2 A car accelerates uniformly from $10 \, \text{m s}^{-1}$ to $20 \, \text{m s}^{-1}$ as it travels 500 m along a straight road.

   (a) Find the acceleration of the car.

   (b) Find the time taken by the car to travel the 500 m.

3 A car accelerates uniformly as it moves along a straight road. Its velocity increases from $5 \, \text{m s}^{-1}$ to $12 \, \text{m s}^{-1}$ in a 10-second period.

   (a) Find the acceleration of the car.

   (b) Find the distance travelled by the car.

4 A lift rises vertically from rest, accelerating at a constant rate until it reaches a speed of 1.6 m s^{-1} after 8 seconds.

(a) Find the acceleration of the lift.

(b) The lift continues to accelerate for a further 2 seconds. Find the distance that the lift has now risen.

(c) The lift then shows down, at a constant rate, and stops after a further 5 seconds. Find the total distance travelled by the lift.

5 A car accelerates uniformly from a speed of 50 km h^{-1} to a speed of 80 km h^{-1} in 20 seconds as it travels along a straight road.

(a) Calculate the acceleration of the car in m s^{-2}.

(b) Calculate the distance travelled during the 20 seconds, giving your answer to the nearest metre.

6 A van travelling at 40 mph skids to a halt in a distance of 15 m. Find the acceleration of the van and the time taken to stop, assuming that the deceleration is uniform and the van travels along a straight line. (Assume 1 mile = 1600 m.)

7 A train signal is placed so that a train can decelerate uniformly from a speed of 96 km h^{-1} to come to rest at the end of a platform. For passenger comfort the deceleration must be no greater than 0.4 m s^{-2}. Assume that the train travels along a straight line. Calculate:

(a) the shortest distance the signal can be from the end of the platform

(b) the shortest time for the train to decelerate.

8 A rocket is travelling with a velocity of 80 m s^{-1}. The engines are switched on for 6 seconds and the rocket accelerates uniformly at 40 m s^{-2}. Assume that the rocket is always moving in the same direction. Calculate the distance travelled over the 6 seconds.

9 The world record for the men's 60 m race was 6.41 seconds.

(a) Assuming that the race was carried out under constant acceleration, calculate the acceleration of the runner and his speed at the end of the race.

(b) Now assume that in a 100 m race the runner accelerates for the first 60 m and completes the race by running the next 40 m at the speed you calculated in **(a)**.

Calculate the time for the athlete to complete the race.

10 The world record for the men's 100 m sprint was 9.83 seconds. Assume that the last 40 m was run at constant speed and that the acceleration during the first 60 m was constant.

   (a) Calculate this speed.

   (b) Calculate the acceleration of the athlete.

11 Telegraph poles, 40 m apart stand alongside a straight railway line. The times taken for a locomotive to pass the two gaps between three consecutive poles are 2.5 seconds and 2.3 seconds, respectively. Assume that the acceleration of the train is constant. Calculate the acceleration of the train and the speed past the first post.

12 A sprinter starts from rest, and accelerates at 2 m s^{-2} for the first 4 seconds of a race. Assume that the sprinter moves along a straight line.

   (a) Find the distance travelled by the sprinter in the first 4 seconds.

   (b) Find the speed of the sprinter at the end of the first 4 seconds.

   (c) The sprinter then travels at this speed for the remainder of the race. He travels a total distance of 100 metres. Find the total time that he takes to complete the race. [A]

13 A cyclist is travelling along a straight horizontal road. As she passes a bus stop she sees a red traffic light ahead of her. She continues to travel with a constant speed of 3 m s^{-1} for 20 seconds and then decelerates at a constant rate of 0.2 m s^{-2} until coming to rest at the traffic light.

   (a) Calculate the distance between the bus stop and the traffic light.

   (b) Calculate the time the cyclist takes to travel from the bus stop to the traffic light.

   (c) Calculate the average speed of the cyclist between the bus stop and the traffic light. [A]

14 A set of traffic lights covers road repairs on a road. The traffic lights are 80 m apart. Assume that a car accelerates at 2 m s^{-2} from rest until it reaches a speed of 10 m s^{-1} and then travels at a constant speed. What is the least time taken for a car starting from rest at the first set of lights to reach the next set?

15 A train starts from rest and moves along a straight track with constant acceleration $\frac{1}{3}$ m s^{-2} for 2 minutes. For the next 4 minutes the train moves with zero acceleration, after which a uniform retardation of 2 m s^{-2} brings it to rest. Find the total distance travelled by the train from starting to stopping. [A]

16 As a train leaves a station it accelerates, from rest, at 0.8 m s^{-2} for 30 seconds, travels at a constant speed for the next 5 minutes and then slows down, stopping in 20 seconds at a second station. Assume that the train moves along a straight track.

   (a) Find the maximum speed of the train.

   (b) Find the distance travelled by the train between the stations, clearly stating any assumptions that you have made.　　　　　　　　　　　　　　　　　　　　[A]

17 As a lift moves upwards from rest it accelerates at 0.8 m s^{-2} for 2 seconds, then travels 4 m at constant speed and finally slows down, with a constant deceleration, stopping in 3 seconds.

   Find the total distance travelled by the lift and the total time taken.　　　　　　　　　　　　　　　　　　　　[A]

18 Two cars, *A* and *B*, are initially at rest side by side. *A* starts off on a straight track with an acceleration of 2 m s^{-2}. Five seconds later *B* starts off on a parallel track to *A*, with acceleration 3.125 m s^{-2}.

   (a) Calculate the distance travelled by *A* after 5 seconds.

   (b) Calculate the time taken for *B* to catch up with *A*.

   (c) Find the speeds of *A* and *B* at that time.

19 A cyclist sets off from rest and moves along a straight horizontal road until she again comes to rest. The motion of the cyclist can be modelled as **three** separate stages.

   In the first stage she accelerates uniformly from rest for 5 seconds until she reaches a velocity $V \text{ m s}^{-1}$.

   She then moves with constant velocity $V \text{ m s}^{-1}$ for 55 seconds.

   Finally she moves with a constant retardation for 10 seconds until coming to rest.

   (a) Sketch a velocity–time graph to show the motion of the cyclist.

   (b) Given that the total length of the journey is 300 m, find the value of *V*.　　　　　　　　　　　　　　　　　　[A]

20 Two cars are initially 36 m apart travelling in the same direction along a straight, horizontal road. The car in front is initially travelling at 10 m s^{-1}, but decelerating at 2 m s^{-2}. The other car travels at a constant 15 m s^{-1}.

   (a) Model the cars as particles. By finding the distance travelled by each car after *t* seconds, show that the distance between the two cars is $36 - 5t - t^2$ metres. Find when they would collide if neither car takes avoiding action.

   (b) Would it be necessary to revise your answers to part **(a)** if the cars were not modelled as particles? Give reasons to support your answer.　　　　　　　　　　　　　[A]

2.4 Motion under gravity

For many centuries it was believed that:

(a) heavier bodies fell faster than light ones, and

(b) the speed of a falling body was constant all through its motion.

> *Galileo Galilei* (1564–1642) was the first person to state clearly (and to demonstrate) that all objects fall with the same acceleration.

> Modern scientific instruments determine the acceleration of falling bodies as values in the region of 9.81 m s^{-2}, although the value varies slightly at different places on the earth's surface, and at different altitudes. The symbol used to represent this 'acceleration due to gravity' is g.

The value of g is sometimes approximated to 10 m s^{-2}, but in this book we shall normally use 9.8 m s^{-2} unless stated otherwise. Since this acceleration acts towards the earth's surface, its sign must always be opposite to that of any velocities that are upwards.

When solving problems involving motion under gravity (ignoring any air resistance at this stage) the formulae for motion with constant acceleration may be used.

Worked example 2.7

A ball is projected vertically upwards with an initial speed of 30 m s^{-1}. Calculate the maximum height reached.

Solution

At the top of the ball's flight, its velocity will be zero.

Take the upwards direction as positive, so the acceleration will be $a = -g = -9.8\,\text{m s}^{-2}$.

Using $v^2 = u^2 + 2as$ with $v = 0\,\text{m s}^{-1}$, $u = 30\,\text{m s}^{-1}$ and $a = -9.8\,\text{m s}^{-2}$ gives

$$0 = 30^2 - 2 \times 9.8 \times h$$

where h is the maximum height reached.

Thus

$$h = \frac{900}{19.6} = 45.9\,\text{m (to three significant figures).}$$

At the top of the flight the velocity of the ball is zero

h

Worked example 2.8

A stone is fired vertically upwards from a catapult with an initial speed of 10 m s^{-1}. Find the time for which the ball is 5 m above the point of release.

Solution

The stone is at a height of 5 m on its upward journey and again on the way down. The neatest way to solve the problem is as follows.

Using $s = ut + \dfrac{1}{2}at^2$ with $s = 5$ m, $u = 10 \text{ m s}^{-1}$ and $a = -9.8 \text{ m s}^{-2}$ gives

$$5 = 10t - 0.5 \times 9.8 \times t^2$$

where t seconds is the time passed since the stone was thrown up. Hence,

$$4.9t^2 - 10t + 5 = 0$$

Using the quadratic equation formula

$$t = \frac{10 \pm \sqrt{2}}{9.8} = 0.876 \text{ or } 1.165$$

the required time interval is $1.165 - 0.876 = 0.289$ seconds.

EXERCISE 2C

1 A ball is dropped from rest at a height of 2 m. Find the time that the ball takes to fall to the ground, if it is:
   (a) on the Earth,
   (b) on the Moon, where $g = 1.6 \text{ m s}^{-2}$.

2 A ball, that is initially at rest, falls from a height of 3 m to the ground.
   (a) Find the time that the ball takes to fall this distance.
   (b) Find the speed of the ball when it hits the ground.

3 A ball is thrown upwards with an initial speed of 14.7 m s^{-1} from a height of 1 m above ground level.
   (a) Find the time that it takes the ball to reach its maximum height.
   (b) Find the maximum height of the ball above ground level.
   (c) Find the speed of the ball when it hits the ground.

4 A rocket rises vertically from ground level to a height of 100 m in 10 seconds. Assume that the acceleration of the rocket is constant and that it starts at rest.
   (a) Find the acceleration of the rocket and its speed at a height of 100 m.

After these 10 seconds the rocket runs out of fuel, but continues to move vertically under the influence of gravity.

(b) Find the maximum height reached by the rocket.

5 The diagram shows three positions of a ball which has been thrown vertically upwards with a velocity of u m s^{-1}

Position A is the initial position.

Position B is halfway up.

Position C is at the top of the motion.

Copy the diagram and for each position put on arrows where appropriate to show the direction of the velocity.

On the same diagram put on arrows to show the direction of the acceleration.

6 A ball is thrown vertically upwards from ground level with an initial speed of 7 m s^{-1}. Assume that no resistance forces act on the ball, so that it moves only under the influence of gravity.

(a) Find the maximum height of the ball.

(b) The ball hits the ground T seconds after it was thrown. Find T. [A]

7 A stone is released from rest and falls vertically through a distance of 22.5 metres before hitting the ground.

(a) Calculate the velocity of the stone as it hits the ground.

(b) Calculate the time between the stone being released and hitting the ground.

(c) Sketch a velocity–time graph to show the motion of the stone while it is falling.

(d) State one modelling assumption that you have made in order to answer the question. [A]

8 (a) A stone is released from rest and falls vertically through a distance of 10 metres before hitting the ground.

    (i) Calculate the speed at which the stone hits the ground.

    (ii) Calculate the time between the stone being released and hitting the ground.

(b) The same stone is released from a height of 10 metres above the surface of the Moon. On the Moon the acceleration due to gravity is much lower than on the Earth.

If the calculations in **(a)(i)** and **(a)(ii)** were carried out for the motion of the stone on the Moon instead of the motion on the Earth, explain how your answers would be affected. [A]

9 A ball is dropped from rest on to level ground from a height of 20 m.

    (a) Calculate the time taken to reach the ground.

    The ball rebounds with half the speed it strikes the ground.

    (b) Calculate the time taken to reach the ground a second time.

10 A stone is thrown vertically downwards from a high building with an initial velocity of 4 m s^{-1}. Calculate the time required for the stone to travel 30 m and its speed at this time.

11 A ball is thrown vertically upwards from the top of a cliff which is 50 m high. The initial velocity of the ball is 25 m s^{-1}. Calculate the time taken to reach the bottom of the cliff and the speed of the ball at that instant.

12 One stone is thrown vertically upwards with a speed of 2 m s^{-1} and another is thrown vertically downwards with a speed of 2 m s^{-1}. Both are thrown at the same time from a window 5 m above ground level.

    (a) Which stone hits the ground first?

    (b) Which stone is travelling fastest when it hits the ground?

    (c) What is the total distance travelled by each stone?

13 A ball is thrown vertically upwards with an initial velocity of 30 m s^{-1}. One second later, another ball is thrown upwards with an initial velocity of $u \text{ m s}^{-1}$. The balls collide after a further 2 seconds. Find the value of u.

14 When a ball hits the ground it rebounds with half of the speed that it had when it hit the ground. If the ball is dropped from rest, at a height h, calculate the height to which it rebounds.

15 A small canister is attached to a helium-filled balloon and released from rest at ground level. After 4 seconds it is moving vertically upwards at 6 m s^{-1}.

    (a) Find the height of the balloon and canister after 4 seconds, stating clearly any assumptions that you make.

    When the balloon reaches a height of 27 m it bursts.

    (b) Find the maximum height reached by the canister.   [A]

16 A tennis ball is hit so that it moves vertically downwards from a height of 1 m with an initial speed of 5 m s^{-1}. When it hits the ground it rebounds vertically with half the speed it had when it hit the ground.

    (a) Find the height to which it rebounds.

    (b) State whether this is likely to be an under-estimate or an over-estimate, giving reasons to support your answer.   [A]

Key point summary

Formulae to learn:

$v = u + at$

$s = \dfrac{1}{2}(u + v)t$

$v^2 = u^2 + 2as$

$s = ut + \dfrac{1}{2}at^2$

$s = vt - \dfrac{1}{2}at^2$

1 The gradient of a displacement–time graph gives the velocity. *p9*

2 The gradient of a velocity–time graph gives the acceleration. *p10*

3 The area under a velocity–time graph can be used to find the displacement. *p11*

4 Only use the constant acceleration formulae when the acceleration is constant or can be assumed to be constant. *p19*

5 All objects accelerate in the same way when falling under the influence of gravity alone. *p25*

6 The acceleration due to gravity is 9.8 m s^{-2}. *p25*

Test yourself	What to review
1 A car accelerates uniformly, along a straight road, from rest to 20 m s^{-1} in 25 seconds. It travels at this speed for 1.5 minutes and then slows down, stopping after a further 35 seconds. **(a)** Draw a velocity–time graph and use it to find the total distance travelled by the car. **(b)** Calculate the acceleration of the car on each stage of its journey.	*Section 2.2*
2 The velocity of a car increases from 5 m s^{-1} to 25 m s^{-1} as it travels a distance of 100 m. Assume that the acceleration of the car is constant and that the car moves along a straight line. **(a)** Find the acceleration of the car. **(b)** Find the speed of the car when it has travelled 50 m. **(c)** Find the time it takes for the car to travel the 100 m.	*Section 2.3*

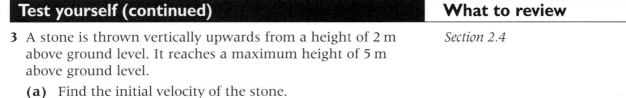

Test yourself (continued)   What to review

3 A stone is thrown vertically upwards from a height of 2 m   *Section 2.4*
above ground level. It reaches a maximum height of 5 m
above ground level.

   (a) Find the initial velocity of the stone.
   (b) Find the velocity of the stone when it hits the ground.
   (c) How long is the stone in the air?

Test yourself   ANSWERS

3 (a) 7.67 m s^{-1};   **(b)** 9.90 m s^{-1};   **(c)** 1.79 s.

2 (a) 3 m s^{-2};   **(b)** 18.0 m s^{-1};   **(c)** $6\frac{2}{3}$ s.

   (b) 0.8 m s^{-2}, 0 m s^{-2}, $-\frac{4}{7}$ m s^{-2}.

   2400 m;

1 (a)

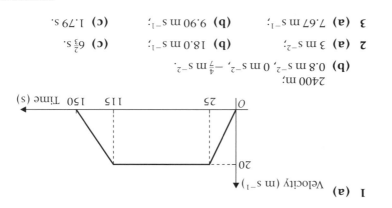

CHAPTER 3
Kinematics in two dimensions

Learning objectives

After studying this chapter you should be able to:
- plot and interpret paths given a position vector
- write positions, velocities and accelerations in the form $x\mathbf{i} + y\mathbf{j}$
- find magnitudes and directions of vectors
- apply and use the constant acceleration equations in two dimensions.
- use velocity triangles to solve simple problems.

3.1 Introduction

In this chapter the idea of using constant acceleration is extended into two dimensions. Before doing this, we need to be able to describe motion in more than one dimension. This is done using vectors, which we will consider first in this chapter.

3.2 Describing motion in two dimensions

We will now consider how to describe motion in two dimensions.

The first key step is to define an origin or reference point. The position of an object is then described relative to this point. Often the origin will be the initial position of an object.

Secondly we define two perpendicular unit vectors. These have length and are directed at right angles to each other.

The diagram at the top of the next page shows an origin O and two unit vectors $\mathbf{i}$ and $\mathbf{j}$.

In addition the diagram shows some points. We can describe the position of each point by using the unit vectors $\mathbf{i}$ and $\mathbf{j}$. For example the position of the point A can be written as:

$$\mathbf{r}_A = 4\mathbf{i} + 3\mathbf{j}$$

Similarly,

$$\mathbf{r}_B = 5\mathbf{i} - 2\mathbf{j}$$
$$\mathbf{r}_C = -5\mathbf{i} - 3\mathbf{j}$$
$$\mathbf{r}_D = -3\mathbf{i} + 2\mathbf{j}$$

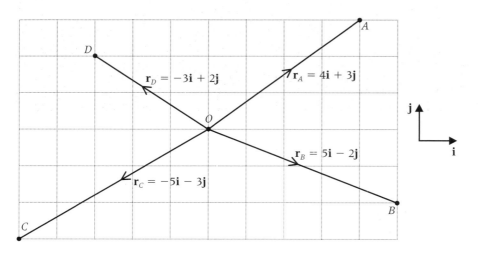

An alternative notation is to use column vectors. Using column vectors these position vectors would be

$$\mathbf{r}_A = \begin{bmatrix} 4 \\ 3 \end{bmatrix}, \mathbf{r}_B = \begin{bmatrix} 5 \\ -2 \end{bmatrix}, \mathbf{r}_C = \begin{bmatrix} -5 \\ -3 \end{bmatrix} \text{ and } \mathbf{r}_D = \begin{bmatrix} -3 \\ 2 \end{bmatrix}.$$

The position vector $\mathbf{r} = x\mathbf{i} + y\mathbf{j}$, could be written as $\begin{bmatrix} x \\ y \end{bmatrix}$.

It is possible to describe the position of a particle in terms of the time, t, that it has been moving. For example we could write:

$$\mathbf{r} = (4t + 3)\mathbf{i} + (t^2 - 4t)\mathbf{j}$$

Then the position vector $\mathbf{r}$ can be found for any value of t and the path of the object can be drawn or described.

For two-dimensional motion we usually write:

$$\mathbf{r} = x(t)\mathbf{i} + y(t)\mathbf{j} \quad \text{or} \quad \mathbf{r} = \begin{bmatrix} x(t) \\ y(t) \end{bmatrix}$$

where $\mathbf{i}$ and $\mathbf{j}$ lie in the plane that the motion takes place in. If this is a vertical plane it is the normal convention that $\mathbf{i}$ is horizontal and $\mathbf{j}$ is vertical.

Then the position vector $\mathbf{r}$ can be found for any value of t and the path of the object can be drawn or described.

Worked example 3.1

A ball is thrown so that its position, in metres, at time t seconds is given by

$$\mathbf{r} = 6t\mathbf{i} + (1 + 8t - 5t^2)\mathbf{j}$$

where $\mathbf{i}$ and $\mathbf{j}$ are horizontal and vertical unit vectors, respectively.

(a) Find the position of the particle when $t = 0, 0.5, 1, 1.5$ and 2 seconds.

(b) Plot the positions in **(a)** and draw the path of the particle.

Solution

(a) The table below shows how the values of t are substituted.

t	$\mathbf{r}$
0	$6 \times 0\mathbf{i} + (1 + 8 \times 0 - 5 \times 0^2)\mathbf{j} = 0\mathbf{i} + 1\mathbf{j}$
0.5	$6 \times 0.5\mathbf{i} + (1 + 8 \times 0.5 - 5 \times 0.5^2)\mathbf{j} = 3\mathbf{i} + 3.75\mathbf{j}$
1	$6 \times 1\mathbf{i} + (1 + 8 \times 1 - 5 \times 1^2)\mathbf{j} = 6\mathbf{i} + 4\mathbf{j}$
1.5	$6 \times 1.5\mathbf{i} + (1 + 8 \times 1.5 - 5 \times 1.5^2)\mathbf{j} = 9\mathbf{i} + 1.75\mathbf{j}$
2	$6 \times 2\mathbf{i} + (1 + 8 \times 2 - 5 \times 2^2)\mathbf{j} = 12\mathbf{i} - 3\mathbf{j}$

(b) These positions can then be plotted and a curve drawn to show the path of the ball.

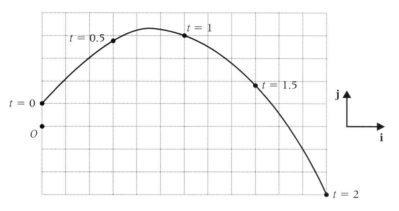

If the origin is at ground level we can note, from the diagram, that:

(i) the ball is thrown from a height of 1 m

(ii) the ball reaches a maximum height of 4.2 m

(iii) the horizontal distance travelled by the ball until it hits the ground is 10.3 m.

Worked example 3.2

Two boats move on a pond. They both start at the same place. One follows a curved path and the other travels along a straight line. The positions, in metres, of the boats, at time t seconds are given by:

$$\mathbf{r}_A = 2t\mathbf{i} + 4t\mathbf{j}$$
$$\mathbf{r}_B = (8t - t^2)\mathbf{i} + 4t\mathbf{j}$$

where $\mathbf{i}$ and $\mathbf{j}$ are unit vectors directed east and north, respectively.

(a) Find the positions of the boats when $t = 0, 2, 4$ and 6 seconds.

(b) What happens when $t = 6$?

(c) Plot the paths of the two boats.

Solution

(a) The table below shows how the positions are calculated by substituting the required values of t.

t	$\mathbf{r}_A$	$\mathbf{r}_B$
0	$2 \times 0\mathbf{i} + 4 \times 0\mathbf{j} = 0\mathbf{i} + 0\mathbf{j}$	$(8 \times 0 - 0^2)\mathbf{i} + 4 \times 0\mathbf{j} = 0\mathbf{i} + 0\mathbf{j}$
2	$2 \times 2\mathbf{i} + 4 \times 2\mathbf{j} = 4\mathbf{i} + 8\mathbf{j}$	$(8 \times 2 - 2^2)\mathbf{i} + 4 \times 2\mathbf{j} = 12\mathbf{i} + 8\mathbf{j}$
4	$2 \times 4\mathbf{i} + 4 \times 4\mathbf{j} = 8\mathbf{i} + 16\mathbf{j}$	$(8 \times 4 - 4^2)\mathbf{i} + 4 \times 4\mathbf{j} = 16\mathbf{i} + 16\mathbf{j}$
6	$2 \times 6\mathbf{i} + 4 \times 6\mathbf{j} = 12\mathbf{i} + 24\mathbf{j}$	$(8 \times 6 - 6^2)\mathbf{i} + 4 \times 6\mathbf{j} = 12\mathbf{i} + 24\mathbf{j}$

(b) When $t = 6$ the boats both have the same position at the same time and so will collide.

(c) The paths of the boats are shown below.

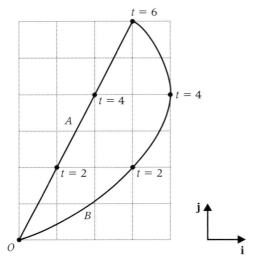

Worked example 3.3

A boat moves so that its position, in metres, at time t seconds is given by

$$\mathbf{r} = (450 - 5t)\mathbf{i} + (4t + 100)\mathbf{j}$$

where $\mathbf{i}$ and $\mathbf{j}$ are unit vectors that are directed east and north, respectively. A rock has position $250\mathbf{i} + 200\mathbf{j}$.

(a) Calculate the time when the boat is due north of the rock.

(b) Calculate the time when the boat is due east of the rock.

Solution

(a) When the boat is due north of the rock, its position will be $250\mathbf{i} + y\mathbf{j}$, where y is a constant. Equating the $\mathbf{i}$ components of the position vectors gives

$$250 = 450 - 5t$$

$$t = \frac{200}{5} = 40 \text{ seconds}$$

At this time the position of the boat is $250\mathbf{i} + 260\mathbf{j}$.

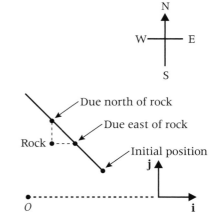

(b) When the boat is due east of the rock, its position will be $x\mathbf{i} + 200\mathbf{j}$, where x is a constant. Equating the $\mathbf{j}$ components of the position vectors gives

$$200 = 100 + 4t$$

$$t = \frac{100}{4} = 25 \text{ seconds}$$

At this time the position of the boat will be $325\mathbf{i} + 200\mathbf{j}$.

EXERCISE 3A

1 A golf ball is hit so that its position, in metres, at time t seconds is given by:

$$\mathbf{r} = 30t\mathbf{i} + (25t - 4.9t^2)\mathbf{j}$$

where $\mathbf{i}$ and $\mathbf{j}$ are horizontal and vertical unit vectors, respectively.

(a) Find the position of the ball when $t = 0$, 1, 2, 3, 4, 5 and 6 seconds.

(b) Plot the path of the ball.

(c) From your plot estimate the horizontal distance travelled by the ball when it hits the ground.

2 A ball moves so that its position vector, in metres, at time t seconds is given by:

$$\mathbf{r} = 5t\mathbf{i} + (1 + 4t - 5t^2)\mathbf{j}$$

where $\mathbf{i}$ and $\mathbf{j}$ are horizontal and vertical unit vectors, respectively.

(a) Find the position of the ball when $t = 0$, $t = 0.2$, $t = 0.4$, $t = 0.6$ and $t = 1$.

(b) Use your answers to **(a)** to sketch the path of the ball.

3 Two children, A and B, run so that their position vectors in metres at time t seconds are given by:

$$\mathbf{r}_A = t\mathbf{i} + t\mathbf{j} \quad \text{and} \quad \mathbf{r}_B = (4 + 4t - t^2)\mathbf{i} + t\mathbf{j}$$

where $\mathbf{i}$ and $\mathbf{j}$ are unit vectors directed east and north, respectively.

Plot the paths of the two children for $0 \le t \le 4$. What happens when $t = 4$?

4 A boat moves so that its position, in metres, at time t seconds is given by:

$$\mathbf{r} = (2t + 6)\mathbf{i} + (3t - 9)\mathbf{j}$$

where $\mathbf{i}$ and $\mathbf{j}$ are unit vectors that are directed east and north respectively. A lighthouse has position $186\mathbf{i} + 281\mathbf{j}$.

(a) Find the position of the boat, when $t = 0$, 40, 80 and 120 seconds.

(b) Plot the path of the boat.

(c) Calculate the time when the boat is due south of the lighthouse and the time when the boat is due east of the lighthouse.

5 A bullet is fired from a rifle, so that its position, in metres, at time t seconds is given by:

$$\mathbf{r} = 180t\mathbf{i} + (1.225 - 4.9t^2)\mathbf{j}$$

where $\mathbf{i}$ and $\mathbf{j}$ are horizontal and vertical unit vectors, respectively, and the origin is at ground level.

(a) Find the initial height of the bullet.

(b) Find the time when the bullet hits the ground.

(c) Find the horizontal distance travelled by the bullet.

6 A boomerang is thrown. As it moves its position, in metres, at time t seconds is modelled by:

$$\mathbf{r} = (10t - t^2)\mathbf{i} + (2 + t - t^2)\mathbf{j}$$

where $\mathbf{i}$ and $\mathbf{j}$ are horizontal and vertical unit vectors. The origin is at ground level.

Find the position of the boomerang when it hits the ground.

3.3 Expressing quantities as vectors

Not all quantities such as positions, velocities or accelerations will be expressed in the form $a\mathbf{i} + b\mathbf{j}$. For example a position may be expressed in terms of a distance and a bearing. In this section we see how to express these quantities in this form.

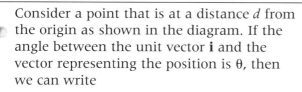

Consider a point that is at a distance d from the origin as shown in the diagram. If the angle between the unit vector $\mathbf{i}$ and the vector representing the position is θ, then we can write

$$\mathbf{r} = d\cos\theta\,\mathbf{i} + d\sin\theta\,\mathbf{j}.$$

If $\mathbf{i}$ and $\mathbf{j}$ are horizontal and vertical, respectively, we would say that the horizontal component of the position vector is $d\cos\theta$ and that the vertical component is $d\sin\theta$.

As a column vector we would write

$$\mathbf{r} = \begin{bmatrix} d\cos\theta \\ d\sin\theta \end{bmatrix}.$$

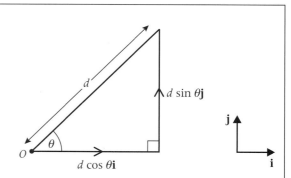

The following examples will illustrate how to write quantities in the form $a\mathbf{i} + b\mathbf{j}$ or as column vectors.

Worked example 3.4

A ship starts at the origin and travels 200 km on a bearing of 140°. Express the position of the ship in the form $a\mathbf{i} + b\mathbf{j}$, where $\mathbf{i}$ and $\mathbf{j}$ are unit vectors that are directed east and north, respectively.

Solution

The diagram shows the position of the ship and the unit vectors.

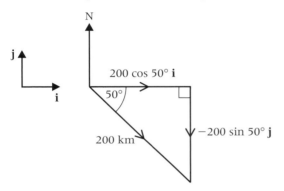

In this case we can write:

$$\mathbf{r} = 200 \cos 50°\mathbf{i} - 200 \sin 50°\mathbf{j}$$

Or as a column vector:

$$\mathbf{r} = \begin{bmatrix} 200 \cos 50° \\ -200 \sin 50° \end{bmatrix}$$

Worked example 3.5

A model aeroplane is travelling at 15 m s^{-1} on a bearing of 250°. Express the velocity of the aeroplane in the form $a\mathbf{i} + b\mathbf{j}$, where $\mathbf{i}$ and $\mathbf{j}$ are unit vectors that are directed east and north, respectively.

Solution

The diagram shows the velocity of the aeroplane and the unit vectors $\mathbf{i}$ and $\mathbf{j}$. In this case the velocity of the aeroplane can be expressed as

$$\mathbf{v} = -15 \cos 20°\mathbf{i} - 15 \sin 20°\mathbf{j}$$

Or as a column vector:

$$\mathbf{v} = \begin{bmatrix} -15 \cos 20° \\ -15 \sin 20° \end{bmatrix}.$$

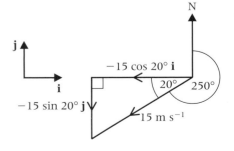

Worked example 3.6

The velocity of a bird is $3\mathbf{i} + 4\mathbf{j}$, where $\mathbf{i}$ and $\mathbf{j}$ are unit vectors directed east and north, respectively. Find the speed of the bird and the direction in which it is heading.

Solution

The diagram shows the velocity of the bird and the unit vectors. The speed of the bird is given by the magnitude or length of the velocity vector. This can be calculated using Pythagoras' theorem.

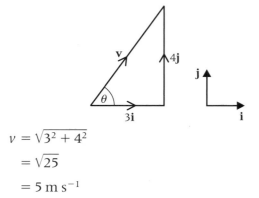

$$v = \sqrt{3^2 + 4^2}$$
$$= \sqrt{25}$$
$$= 5 \text{ m s}^{-1}$$

Next find the angle θ marked on the diagram.

$$\tan \theta = \frac{4}{3}$$
$$\theta = 53.1°$$

The best way to describe the direction of the velocity of the bird is by using the bearing of the direction that it is heading in. This is given by

$$90 - 53.1 = 036.9°$$

Note. The acceleration due to gravity is often expressed as $-g\mathbf{j}$, because it has magnitude g m s^{-2} and acts vertically downwards.

Working with vectors

In this chapter you will need to be able to add and subtract vectors, and to multiply a vector by a scalar. The key results that you will require are shown below.

> If $\quad \mathbf{p} = a\mathbf{i} + b\mathbf{j}$ and $\quad \mathbf{q} = c\mathbf{i} + d\mathbf{j}$,
> then $\mathbf{p} + \mathbf{q} = (a + c)\mathbf{i} + (b + d)\mathbf{j}$
> and $\quad \mathbf{p} - \mathbf{q} = (a - c)\mathbf{i} + (b - d)\mathbf{j}$.
>
> Also, if k is a scalar,
> then $k\mathbf{p} = ka\mathbf{i} + kb\mathbf{j}$.

> Using column vectors where
> $$\mathbf{p} = \begin{bmatrix} a \\ b \end{bmatrix} \quad \text{and} \quad \mathbf{q} = \begin{bmatrix} c \\ d \end{bmatrix}$$
> so that
> $$\mathbf{p} + \mathbf{q} = \begin{bmatrix} a+c \\ b+d \end{bmatrix}, \mathbf{p} - \mathbf{q} = \begin{bmatrix} a-c \\ b-d \end{bmatrix}$$
> and
> $$k\mathbf{p} = \begin{bmatrix} ka \\ kb \end{bmatrix}.$$

3

Worked example 3.7

If $\mathbf{a} = 4\mathbf{i} + 2\mathbf{j}$ and $\mathbf{b} = 7\mathbf{i} - 3\mathbf{j}$ find

(a) $\mathbf{a} + \mathbf{b}$,   (b) $4\mathbf{a}$,   (c) $3\mathbf{a} - 2\mathbf{b}$.

Solution

(a) $\begin{aligned}[t] \mathbf{a} + \mathbf{b} &= (4\mathbf{i} + 2\mathbf{j}) + (7\mathbf{i} - 3\mathbf{j}) \\ &= (4 + 7)\mathbf{i} + (2 - 3)\mathbf{j} \\ &= 11\mathbf{i} - \mathbf{j} \end{aligned}$

(b) $\begin{aligned}[t] 4\mathbf{a} &= 4(4\mathbf{i} + 2\mathbf{j}) \\ &= 4 \times 4\mathbf{i} + 4 \times 2\mathbf{j} \\ &= 16\mathbf{i} + 8\mathbf{j} \end{aligned}$

(c) $\begin{aligned}[t] 3\mathbf{a} - 2\mathbf{b} &= 3(4\mathbf{i} + 2\mathbf{j}) - 2(7\mathbf{i} - 3\mathbf{j}) \\ &= (12\mathbf{i} + 6\mathbf{j}) - (14\mathbf{i} - 6\mathbf{j}) \\ &= (12 - 14)\mathbf{i} + (6 + 6)\mathbf{j} \\ &= -2\mathbf{i} + 12\mathbf{j} \end{aligned}$

Worked example 3.8

If $\mathbf{a} = \begin{bmatrix} 4 \\ 7 \end{bmatrix}$ and $\mathbf{b} = \begin{bmatrix} -1 \\ 4 \end{bmatrix}$, find

(a) $\mathbf{a} + 2\mathbf{b}$,   (b) $3\mathbf{a} - 4\mathbf{b}$

Solution

(a) $\begin{aligned}[t] \mathbf{a} + 2\mathbf{b} &= \begin{bmatrix} 4 \\ 7 \end{bmatrix} + 2\begin{bmatrix} -1 \\ 4 \end{bmatrix} \\[2mm] &= \begin{bmatrix} 4 \\ 7 \end{bmatrix} + \begin{bmatrix} -2 \\ 8 \end{bmatrix} \\[2mm] &= \begin{bmatrix} 4-2 \\ 7+8 \end{bmatrix} \\[2mm] &= \begin{bmatrix} 2 \\ 15 \end{bmatrix} \end{aligned}$

(b) $3\mathbf{a} - 4\mathbf{b} = 3\begin{bmatrix} 4 \\ 7 \end{bmatrix} - 4\begin{bmatrix} -1 \\ 4 \end{bmatrix}$

$$= \begin{bmatrix} 12 \\ 21 \end{bmatrix} - \begin{bmatrix} -4 \\ 16 \end{bmatrix}$$

$$= \begin{bmatrix} 12 - (-4) \\ 21 - 16 \end{bmatrix}$$

$$= \begin{bmatrix} 16 \\ 5 \end{bmatrix}$$

EXERCISE 3B

1 The unit vectors **i** and **j** are directed east and north respectively. The positions below are given in terms of a bearing and a distance. Express each position in the form $a\mathbf{i} + b\mathbf{j}$ and as a column vector:

    (a) 45 m on a bearing of 080°,

    (b) 105 m on a bearing of 060°,

    (c) 21 m on a bearing of 340°,

    (d) 62 m on a bearing of 260°,

    (e) 290 m on a bearing of 162°.

2 A boat sails south west at 6 m s^{-1}. Find the velocity of the ship in the form $a\mathbf{i} + b\mathbf{j}$, where **i** and **j** are unit vectors directed east and north, respectively.

3 A ship travels at a speed of 5 m s^{-1}. Express its velocity in terms of the unit vectors **i** and **j**, that are directed east and north, respectively, if the ship is sailing:

    (a) due east,

    (b) due south,

    (c) due west,

    (d) south east,

    (e) north west.

4 A ball is thrown so that its initial velocity is 8 m s^{-1} at an angle of 50° above the horizontal. The unit vectors **i** and **j** are horizontal and vertical, respectively. Express the initial velocity of the ball in terms of the unit vectors **i** and **j**.

5 For each velocity listed below, find its magnitude and direction. The unit vectors **i** and **j** are directed east and north, respectively. Give the directions as the bearing along which the velocity is directed.

    (a) $(4\mathbf{i} + 7\mathbf{j})$ m s^{-1}

    (b) $(5\mathbf{i} - 6\mathbf{j})$ m s^{-1}

    (c) $(-8\mathbf{i} - 9\mathbf{j})$ m s^{-1}

    (d) $(-12\mathbf{i} + 8\mathbf{j})$ m s^{-1}

6 The following velocities are in metres per second. Find the magnitude of each velocity and find its direction, giving your answer as a bearing. The direction of the vector $\begin{bmatrix} 0 \\ 1 \end{bmatrix}$ is due north.

(a) $\begin{bmatrix} 4 \\ 9 \end{bmatrix}$,   (b) $\begin{bmatrix} 2 \\ -3 \end{bmatrix}$,   (c) $\begin{bmatrix} -4 \\ 2 \end{bmatrix}$,   (d) $\begin{bmatrix} -8 \\ -7 \end{bmatrix}$.

7 If $\mathbf{a} = 3\mathbf{i} + 2\mathbf{j}$, $\mathbf{b} = 2\mathbf{i} - 8\mathbf{j}$ and $\mathbf{c} = -4\mathbf{i} + 3\mathbf{j}$, find:

(a) $\mathbf{a} + \mathbf{b}$,   (b) $\mathbf{a} + \mathbf{c}$,   (c) $\mathbf{a} - \mathbf{b}$,
(d) $\mathbf{a} - \mathbf{c}$,   (e) $3\mathbf{a} + 2\mathbf{b}$,   (f) $4\mathbf{b} - 6\mathbf{c}$.

8 If $\mathbf{p} = \begin{bmatrix} 1 \\ -2 \end{bmatrix}$, $\mathbf{q} = \begin{bmatrix} 3 \\ 1 \end{bmatrix}$ and $\mathbf{r} = \begin{bmatrix} -4 \\ 3 \end{bmatrix}$, find:

(a) $\mathbf{p} + \mathbf{q}$,   (b) $\mathbf{p} - \mathbf{q}$,   (c) $\mathbf{r} - \mathbf{p}$,
(d) $3\mathbf{r} + 2\mathbf{q}$,   (e) $8\mathbf{p} - 2\mathbf{q}$,   (f) $3\mathbf{q} - 7\mathbf{p}$.

9 An object moves along a straight line from the point with position vector $(5\mathbf{i} + 6\mathbf{j})$ m to the point with position vector $(8\mathbf{i} - 2\mathbf{j})$ m, where the unit vectors $\mathbf{i}$ and $\mathbf{j}$ are directed east and north, respectively.

(a) Find the distance travelled by the object.

(b) Find the bearing along which the object was travelling.

10 The acceleration of a particle is $(-3\mathbf{i} + 2\mathbf{j})$ m s^{-2}, where the unit vectors $\mathbf{i}$ and $\mathbf{j}$ are directed north and east, respectively. Find the magnitude of the acceleration. Calculate the bearing along which the acceleration is directed.

3.4 Constant acceleration equations in two dimensions

The constant acceleration equations that were introduced and used in the last chapter can be extended into two dimensions as vector equations.

To make use of these vector equations the velocities, accelerations and positions must be in the form $a\mathbf{i} + b\mathbf{j}$ for two-dimensional motion.

The constant acceleration equations become

$$\mathbf{r} = \mathbf{u}t + \frac{1}{2}\mathbf{a}t^2 \quad \text{or} \quad \mathbf{r} = \mathbf{u}t + \frac{1}{2}\mathbf{a}t^2 + \mathbf{r}_0$$

$$\mathbf{v} = \mathbf{u} + \mathbf{a}t$$

$$\mathbf{r} = \frac{1}{2}(\mathbf{u} + \mathbf{v})t \quad \text{or} \quad \mathbf{r} = \frac{1}{2}(\mathbf{u} + \mathbf{v})t + \mathbf{r}_0$$

where $\mathbf{r}$ is the position at time t, $\mathbf{u}$ is the initial velocity, $\mathbf{v}$ is the velocity at time t, $\mathbf{a}$ is the acceleration and $\mathbf{r}_0$ is the initial position.

You will notice the similarity between these equations and the constant acceleration equations that you used in the last chapter. At this stage we will not use a vector equation that is equivalent to $v^2 = u^2 + 2as$.

The following examples illustrate how these formulae can be applied.

Worked example 3.9

A ball is moving on an inclined plane. The initial velocity of the ball is $4\mathbf{i}$ m s^{-1}, its acceleration is $-2\mathbf{j}$ m s^{-2} and its initial position is $(9\mathbf{i} + 4\mathbf{j})$ m, where $\mathbf{i}$ and $\mathbf{j}$ are perpendicular unit vectors that lie in the plane on which the ball is moving.

(a) Find the velocity of the ball after it has been moving for 3 seconds.

(b) Find the position of the ball after it has been moving for 5 seconds.

Solution

For this problem:

$$\mathbf{u} = 4\mathbf{i}$$
$$\mathbf{a} = -2\mathbf{j}$$
$$\mathbf{r}_0 = 9\mathbf{i} + 4\mathbf{j}$$

(a) Using the formula for the velocity with the vectors above and $t = 3$ gives:

$$\mathbf{v} = \mathbf{u} + \mathbf{a}t$$
$$= 4\mathbf{i} + (-2\mathbf{j}) \times 3$$
$$= 4\mathbf{i} - 6\mathbf{j}$$

(b) Using the formula for the position with the vectors listed above and $t = 5$ gives:

$$\mathbf{r} = \mathbf{u}t + \frac{1}{2}\mathbf{a}t^2 + \mathbf{r}_0$$
$$= 4\mathbf{i} \times 5 + \frac{1}{2}(-2\mathbf{j}) \times 5^2 + 9\mathbf{i} + 4\mathbf{j}$$
$$= 20\mathbf{i} - 25\mathbf{j} + 9\mathbf{i} + 4\mathbf{j}$$
$$= 29\mathbf{i} - 21\mathbf{j}$$

Alternative Solution

Working with column vectors gives

$$\mathbf{u} = \begin{bmatrix} 4 \\ 0 \end{bmatrix}, \mathbf{a} = \begin{bmatrix} 0 \\ -2 \end{bmatrix} \text{ and } \mathbf{r}_0 = \begin{bmatrix} 9 \\ 4 \end{bmatrix}.$$

(a) Using $\mathbf{v} = \mathbf{u} + \mathbf{a}t$ with $t = 3$ gives:

$$\mathbf{v} = \begin{bmatrix} 4 \\ 0 \end{bmatrix} + 3\begin{bmatrix} 0 \\ -2 \end{bmatrix}$$

$$= \begin{bmatrix} 4 \\ 0 \end{bmatrix} + \begin{bmatrix} 0 \\ -6 \end{bmatrix}$$

$$= \begin{bmatrix} 4 \\ -6 \end{bmatrix}$$

(b) Using $\mathbf{r} = \mathbf{u}t + \frac{1}{2}\mathbf{a}t^2 + \mathbf{r}_0$ gives

$$\mathbf{r} = 5\begin{bmatrix} 4 \\ 0 \end{bmatrix} + \frac{1}{2} \times 5^2 \begin{bmatrix} 0 \\ -2 \end{bmatrix} + \begin{bmatrix} 9 \\ 4 \end{bmatrix}$$

$$= \begin{bmatrix} 20 \\ 0 \end{bmatrix} + \begin{bmatrix} 0 \\ -25 \end{bmatrix} + \begin{bmatrix} 9 \\ 4 \end{bmatrix}$$

$$= \begin{bmatrix} 29 \\ -21 \end{bmatrix}$$

Worked example 3.10

A boat has initial velocity $(5\mathbf{i} + 3\mathbf{j})$ m s^{-1}. It accelerates for 4 seconds. Its velocity is then $(2\mathbf{i} - 4\mathbf{j})$ m s^{-1}. The boat then stops accelerating and travels with this velocity for a further 20 seconds. The initial position of the boat is $(24\mathbf{i} + 32\mathbf{j})$ m. The unit vectors $\mathbf{i}$ and $\mathbf{j}$ are directed east and north, respectively.

(a) Find the acceleration of the boat.

(b) Find the position of the boat after 4 seconds.

(c) Find the final position of the boat.

Solution

(a) The acceleration can be found using the equation $\mathbf{v} = \mathbf{u} + \mathbf{a}t$. Using $\mathbf{u} = 5\mathbf{i} + 3\mathbf{j}$, $\mathbf{v} = 2\mathbf{i} - 4\mathbf{j}$ and $t = 4$ gives:

$$\mathbf{v} = \mathbf{u} + \mathbf{a}t$$
$$2\mathbf{i} - 4\mathbf{j} = 5\mathbf{i} + 3\mathbf{j} + 4\mathbf{a}$$
$$4\mathbf{a} = -3\mathbf{i} - 7\mathbf{j}$$
$$\mathbf{a} = -\frac{3}{4}\mathbf{i} - \frac{7}{4}\mathbf{j}$$

(b) The position after 4 seconds can be found using the constant acceleration equation $\mathbf{r} = \frac{1}{2}(\mathbf{u} + \mathbf{v})t + \mathbf{r}_0$, with $\mathbf{u} = 5\mathbf{i} + 3\mathbf{j}$, $\mathbf{v} = 2\mathbf{i} - 4\mathbf{j}$, $\mathbf{r}_0 = 24\mathbf{i} + 32\mathbf{j}$ and $t = 4$.

$$\mathbf{r} = \frac{1}{2}(\mathbf{u} + \mathbf{v})t + \mathbf{r}_0$$
$$= \frac{1}{2}(5\mathbf{i} + 3\mathbf{j} + 2\mathbf{i} - 4\mathbf{j}) \times 4 + 24\mathbf{i} + 32\mathbf{j}$$
$$= 14\mathbf{i} - 2\mathbf{j} + 24\mathbf{i} + 32\mathbf{j}$$
$$= 38\mathbf{i} + 30\mathbf{j}$$

(c) After 4 seconds the boat moves with a constant velocity for a further 20 seconds. As the acceleration is zero the equation $\mathbf{r} = \mathbf{u}t + \frac{1}{2}\mathbf{a}t^2 + \mathbf{r}_0$ reduces to $\mathbf{r} = \mathbf{u}t + \mathbf{r}_0$. This can be applied by considering only the motion after the boat stops accelerating. Using $\mathbf{u} = 2\mathbf{i} - 4\mathbf{j}$, $\mathbf{r}_0 = 38\mathbf{i} + 30\mathbf{j}$ and $t = 20$ gives

$$\mathbf{r} = \mathbf{u}t + \mathbf{r}_0$$
$$= (2\mathbf{i} - 4\mathbf{j}) \times 20 + 38\mathbf{i} + 30\mathbf{j}$$
$$= 78\mathbf{i} - 50\mathbf{j}$$

Worked example 3.11

An aeroplane has a constant velocity of $120\mathbf{i}$ m s^{-1}, as it is moving along a runway. It then experiences an acceleration of $(2\mathbf{i} + 5\mathbf{j})$ m s^{-2} for the first 20 seconds of its flight. The unit vectors $\mathbf{i}$ and $\mathbf{j}$ are directed horizontally and vertically, respectively. Assume that the aeroplane is at the origin when it begins to accelerate.

(a) Find an expression for the position of the aeroplane at time t seconds, after it starts to accelerate.

(b) Find the speed of the aeroplane when it is at a height of 250 m.

Solution

(a) As the aeroplane is initially at the origin we can use the constant acceleration equation $\mathbf{r} = \mathbf{u}t + \frac{1}{2}\mathbf{a}t^2$, with $\mathbf{u} = 120\mathbf{i}$ and $\mathbf{a} = 2\mathbf{i} + 5\mathbf{j}$.

$$\mathbf{r} = \mathbf{u}t + \frac{1}{2}\mathbf{a}t^2$$
$$= 120t\mathbf{i} + \frac{1}{2}(2\mathbf{i} + 5\mathbf{j})t^2$$
$$= (120t + t^2)\mathbf{i} + \frac{5}{2}t^2\mathbf{j}$$

(b) The height of the aeroplane is given by the vertical or $\mathbf{j}$ component of the position vector. When the height of the aeroplane is 250 m we have:

$$\frac{5}{2}t^2 = 250$$
$$t^2 = 100$$
$$t = 10 \text{ seconds}$$

The velocity can now be found using this value for t.

$$\mathbf{v} = \mathbf{u} + \mathbf{a}t$$
$$= 120\mathbf{i} + (2\mathbf{i} + 5\mathbf{j}) \times 10$$
$$= 140\mathbf{i} + 50\mathbf{j}$$

The speed can be found as it will be the magnitude of the velocity.

$$v = \sqrt{140^2 + 50^2}$$
$$= \sqrt{22100}$$
$$= 149 \text{ m s}^{-1} \text{ (correct to three significant figures)}$$

EXERCISE 3C

1 During a 10 second period the velocity of a boat changes from $(4\mathbf{i} + 2\mathbf{j})$ m s^{-1} to $(\mathbf{i} - 3\mathbf{j})$ m s^{-1}, where $\mathbf{i}$ and $\mathbf{j}$ are perpendicular unit vectors. Find the acceleration of the boat during this time, assuming that it is constant.

2 The acceleration of a motor boat is $(0.6\mathbf{i} + 0.8\mathbf{j})$ m s^{-2}. Its initial velocity is $3\mathbf{i}$ m s^{-1} and its initial position is $(20\mathbf{i} + 5\mathbf{j})$ m. The unit vectors, $\mathbf{i}$ and $\mathbf{j}$ are directed east and north, respectively.

   (a) Find the velocity of the boat when $t = 3$ seconds.

   (b) Find the position of the boat when $t = 3$ seconds.

3 An object has initial velocity $(3\mathbf{i} - 5\mathbf{j})$ m s^{-1} and an acceleration of $(\mathbf{i} + \mathbf{j})$ m s^{-2}, where $\mathbf{i}$ and $\mathbf{j}$ are perpendicular unit vectors. If it starts at the origin find the position and velocity of the object at time t seconds.

4 The acceleration of a body is $(6\mathbf{i} + 8\mathbf{j})$ m s^{-2}. If the body starts at rest at $0\mathbf{i} + 0\mathbf{j}$ and accelerates for 6 seconds find the velocity and position of the body after 6 seconds. The unit vectors $\mathbf{i}$ and $\mathbf{j}$ are perpendicular.

5 A ball has initial position $2\mathbf{j}$ m, initial velocity $(4\mathbf{i} + 9\mathbf{j})$ m s^{-1} and acceleration $-10\mathbf{j}$ m s^{-2}, where $\mathbf{i}$ and $\mathbf{j}$ are horizontal and vertical unit vectors, respectively. Find the position of the ball at time t seconds and its position when it hits the ground, that is when the vertical component of its position is zero.

6 A snooker ball is struck so that it moves with the constant velocity shown in the diagram. It starts at the point with position vector $(0.2\mathbf{i} + 0.1\mathbf{j})$ m relative to the origin O. Find an expression for the position of the ball at time t, and find where it first hits a cushion. The unit vectors $\mathbf{i}$ and $\mathbf{j}$ are directed as shown on the diagram.

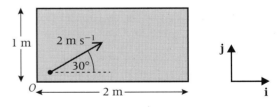

7 A ball is launched from a point with position $(0\mathbf{i} + 1.9\mathbf{j})$ m and velocity $(7\mathbf{i} + 32\mathbf{j})$ m s^{-1}. The ball experiences an acceleration of $(-9.8\mathbf{j})$ m s^{-2} while in the air. The unit vectors $\mathbf{i}$ and $\mathbf{j}$ are horizontal and vertical, respectively.

   (a) Show that the position of the ball, in metres, at time t seconds is given by:

$$\mathbf{r} = 7t\mathbf{i} + (1.9 + 32t - 4.9t^2)\mathbf{j}.$$

   (b) Find how long the ball is in the air and where it lands.

   (c) Find the speed of the ball when it hits the ground.

8 A golf ball is hit from ground level so that its initial velocity is $(20\mathbf{i} + 30\mathbf{j})$ m s^{-1} and its acceleration is $-10\mathbf{j}$ m s^{-2}, where $\mathbf{i}$ and $\mathbf{j}$ are horizontal and vertical unit vectors, respectively. Assume that the initial position of the ball is $0\mathbf{i} + 0\mathbf{j}$.

   (a) Find the time that the ball is in the air, the position of the point where it hits the ground and its speed at this time.

   (b) Find the time for which the height of the ball is greater than 25 m.

9 The unit vectors $\mathbf{i}$ and $\mathbf{j}$, are directed north and east, respectively. A boat has initial velocity $(4\mathbf{i} + 6\mathbf{j})$ m s^{-1} and an initial position of $(80\mathbf{i} + 20\mathbf{j})$ m. It experiences an acceleration of $(-0.02\mathbf{i} - 0.04\mathbf{j})$ m s^{-2} for a period of 4 minutes.

   (a) Find an expression for the velocity and position of the particle at time t seconds.

   (b) After 4 minutes the boat stops accelerating and continues with a constant velocity for a further minute before hitting the bank and stopping. Find the position of the point where the boat hits the bank.

   (c) Find the velocity of the boat when its position is $476\mathbf{i} + 452\mathbf{j}$.

10 A model helicopter has initial position $80\mathbf{j}$ m and flies with a constant velocity of $(6\mathbf{i} - 3\mathbf{j})$ m s^{-1}. A model aeroplane flies at the same height as the helicopter. Its initial position is $10\mathbf{j}$ m, its initial velocity is $5\mathbf{i}$ m s^{-1} and its acceleration is $(0.1\mathbf{i} + 0.05\mathbf{j})$ m s^{-2}. The unit vectors $\mathbf{i}$ and $\mathbf{j}$ are directed east and north, respectively.

   (a) Find expressions for the position vectors of the helicopter and the aeroplane at time t seconds.

   (b) The helicopter and the aeroplane collide. Find the time when this collision takes place and the position of the collision.

11 The unit vectors $\mathbf{i}$ and $\mathbf{j}$, are directed east and north, respectively. A boat has initial velocity $(2\mathbf{i} + 3\mathbf{j})$ m s^{-1} and an initial position of $(40\mathbf{i} + 20\mathbf{j})$ m with respect to an origin O. It experiences an acceleration of $(-0.06\mathbf{i} - 0.04\mathbf{j})$ m s^{-2}. Model the boat as a particle.

   (a) Find expressions for the velocity and position, with respect to O, of the boat at time t seconds.

   (b) The boat hits a sand bank when its position is $52\mathbf{i} + 128\mathbf{j}$. Find the value of t when this happens.   [A]

12 A ship moves so that its position vector, in metres, relative to a lighthouse at time t seconds is

$$\mathbf{r} = (80 - 0.4t)\mathbf{i} + (2t - 80)\mathbf{j}$$

where $\mathbf{i}$ and $\mathbf{j}$ are unit vectors directed east and north, respectively.

(a) Find the distance of the ship from the lighthouse when $t = 60$.

(b) Find the times when the ship is:
    (i) due north of the lighthouse,
    (ii) north-east of the lighthouse.   [A]

13 A particle moves in the horizontal plane that contains the perpendicular unit vectors $\mathbf{i}$ and $\mathbf{j}$. Initially it is at the origin and has velocity $18\mathbf{i}$ m s^{-1}. After accelerating for 10 seconds its velocity is $(30\mathbf{i} + 8\mathbf{j})$ m s^{-1}. Assume that the acceleration of the particle is constant.

(a) Find the acceleration of the particle.

(b) Find the position vector of the particle when its velocity is $(36\mathbf{i} + 12\mathbf{j})$ m s^{-1}.   [A]

14 A particle moves with constant acceleration. Initially, the particle is travelling due north at 4 m s^{-1} and 10 seconds later it is travelling east at 6 m s^{-1}. The unit vectors $\mathbf{i}$ and $\mathbf{j}$ are directed east and north respectively.

(a) Write down the initial velocity of the particle as a vector.

(b) Show that the acceleration of the particle is $(0.6\mathbf{i} - 0.4\mathbf{j})$ m s^{-2} and calculate its magnitude.

(c) Assuming that the particle starts at the origin, find its position after 10 seconds.   [A]

15 At time $t = 0$, a boat is travelling due east at a speed of 3 m s^{-1}. The unit vectors $\mathbf{i}$ and $\mathbf{j}$ are directed east and north, respectively.

(a) Write down the initial velocity of the boat in vector form.

(b) The boat has a constant acceleration of $(0.1\mathbf{i} + 0.2\mathbf{j})$ m s^{-2}. Find an expression for the velocity of the boat at time t seconds.

(c) When $t = T$, the boat is travelling north east. Form an equation that T must satisfy, and solve it to show that $T = 30$.

(d) Find the distance of the boat from its initial position when $t = 20$.   [A]

16 A boat moves, with constant acceleration, so that at time t its velocity is given by

$$\mathbf{v} = 2(a - t)\mathbf{i} + 4(3 - t)\mathbf{j}$$

where a is a constant and $\mathbf{i}$ and $\mathbf{j}$ are unit vectors directed east and north, respectively.

(a) The boat is heading due north when $t = 2$. Find a.

(b) Find:

    (i) the initial velocity of the boat,

    (ii) the acceleration of the boat.

(c) Find the distance between the initial position of the boat and its position when $t = 2$.   [A]

17 A boat moves with a constant acceleration of $(0.2\mathbf{i} + 0.1\mathbf{j})$ m s^{-2}, where the unit vectors $\mathbf{i}$ and $\mathbf{j}$ are directed east and north, respectively. Time t is measured in seconds.

At time $t = 10$, the boat is at the point A, which has position vector $(30\mathbf{i} - 35\mathbf{j})$ metres.
At time $t = 20$, the boat is at the point B, which has position vector $(70\mathbf{i} - 40\mathbf{j})$ metres.

(a) Find the vector $\overrightarrow{AB}$.

(b) Find the velocity of the boat when $t = 10$.

(c) Using your answer to **(b)**, find the velocity of the boat when $t = 0$.

(d) Find the position vector of the boat when $t = 0$.   [A]

3.5 Resultant velocities

Very often, the motion of an object is affected by the fact that it is in a medium that is also moving, for example a boat moving in a current.

The resultant velocity can be found by using a vector triangle. A simple case is illustrated in the diagram.

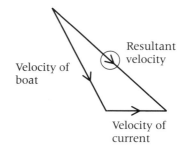

Velocity of boat

Resultant velocity

Velocity of current

Worked example 3.12

A boat would move at 5 m s^{-1} in still water. The boat is in a river in which the current flows at 1 m s^{-1}. If the boat aims to travel at right angles to the bank, find the resultant velocity of the boat, giving:

(a) its magnitude, and

(b) its direction.

Solution

The diagram shows the triangle of velocities for this situation, where v is the resultant speed of the boat.

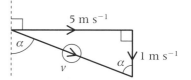

(a) As this is a right-angled triangle,

$$v^2 = 5^2 + 1^2$$
$$v = \sqrt{26}$$
$$= 5.10 \text{ m s}^{-1} \text{ (to 3 significant figures)}$$

(b) The angle α gives the angle between the resultant velocity and the bank.

$$\tan \alpha = \frac{5}{1}$$
$$\alpha = \tan^{-1}(5)$$
$$= 78.7°$$

Worked example 3.13

A boy wants to swim across a river so that his path is at right angles to the bank. He can swim at 3 m s^{-1} in still water and the current flows parallel to the bank at 2 m s^{-1}.

(a) Find the direction in which the boy should swim.

(b) The width of the river is 8 m. Find the time that it would take the boy to swim across the river.

Solution

The diagram shows a triangle of velocities for this problem, where v is the resultant speed.

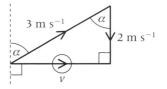

(a) The boy should swim at an angle α to the bank.

$$\cos \alpha = \frac{2}{3}$$
$$\alpha = \cos^{-1}\left(\frac{2}{3}\right)$$
$$= 48.2°$$

(b) As this is a right-angled triangle,

$$3^2 = v^2 + 2^2$$
$$v^2 = 3^2 - 2^2$$
$$v = \sqrt{5} = 2.24 \text{ m s}^{-1} \quad \text{(to 3 significant figures)}$$

The time taken is given by the width divided by the speed.

$$\text{Time taken} = \frac{8}{\sqrt{5}}$$
$$= 3.58 \text{ s (to 3 significant figures).}$$

Worked example 3.14

An aeroplane heads north at 120 m s^{-1} relative to the air. A wind is blowing NE with a constant speed of 50 m s^{-1}.

(a) Find the resultant speed of the aeroplane.

(b) Find the bearing on which the aeroplane travels.

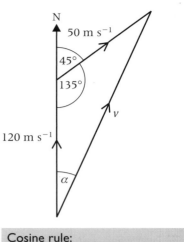

Solution

A triangle of velocities for this problem is shown in the diagram.

(a) Here the cosine rule needs to be used, to find the resultant speed.

$$v^2 = 120^2 + 50^2 - 2 \times 120 \times 50 \cos 135°$$
$$v = 159 \text{ m s}^{-1} \text{ (to 3 significant figures)}$$

(b) Here the sine rule should be used to find the angle α.

$$\frac{\sin \alpha}{50} = \frac{\sin 135°}{v}$$
$$\sin \alpha = \frac{50 \sin 135°}{v} = 0.2219$$
$$\alpha = 12.8°$$

A bearing of $012.8°$.

Cosine rule:
$$a^2 = b^2 + c^2 - 2bc \cos A$$

Sine rule:
$$\frac{\sin A}{a} = \frac{\sin B}{b} = \frac{\sin C}{c}$$

EXERCISE 3D

1 A boat travels at a speed of 6 m s^{-1} in still water. It sets off heading at right angles to the bank of a river. The water in the river is moving at 3 m s^{-1} parallel to the bank.

    (a) Draw a triangle of velocities for this problem.

    (b) Find the angle between the bank and the resultant velocity of the boat.

    (c) Find the resultant speed of the boat.

2 An aeroplane travels at a speed of 90 m s^{-1} relative to the air, and heads due west. A wind blows due north at 40 m s^{-1}.

    (a) Draw a triangle of velocities for this problem.

    (b) Find the resultant speed of the aeroplane.

    (c) Find the bearing of the direction in which the aeroplane actually moves.

3 A girl swims so that she travels across a river at right angles to the bank. A current flows parallel to the bank and has speed 1 m s^{-1}. The girl's resultant velocity is 3 m s^{-1}.

    (a) Find the speed of the girl relative to the water.

    (b) Find the direction in which the girl is heading, giving your answer as the angle between this direction and the bank.

4 A boat is sailing on a bearing of 120° and has a speed of
4 m s⁻¹ relative to the water. A current has speed 2 m s⁻¹ and
flows south west.

   (a) Find the resultant speed of the boat.

   (b) Find the bearing on which the boat moves.

5 An aeroplane wishes to travel due east. A wind is blowing
south east at a speed of 40 m s⁻¹. The speed of the aeroplane
relative to the air is 110 m s⁻¹.

   (a) Find the direction in which the aeroplane should head.

   (b) Find the resultant speed of the aeroplane.

6 A bird tries to fly due north directly from A to B, a distance of
40 m. A wind that blows east causes the bird to move to C
instead of B, arriving 8 seconds after leaving A. Find the
wind's speed.

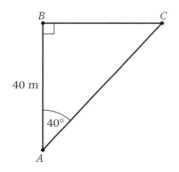

7 A boy is rowing a boat across a river. When he rows in
still water, the boat moves at 2 m s⁻¹. The river flows
parallel to its banks at 1.5 m s⁻¹. The boy aims his boat
perpendicular to the banks of the river, but is carried
downstream by the river.

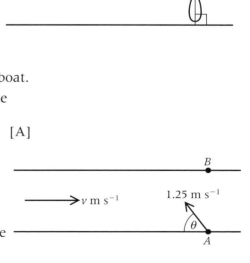

   (a) Sketch an appropriate triangle of velocities.

   (b) Find the magnitude of the resultant velocity of the boat.

   (c) Find the angle this resultant velocity makes with the
   bank of the river, giving your answer to the nearest
   degree.   [A]

8 A girl swims across a river. When she swims in still
water, she swims at 1.25 m s⁻¹. The river flows parallel
to its banks at v m s⁻¹.

   The girl aims to swim upstream at an angle θ degrees to
   the river bank so that her resultant velocity, of magnitude
   1 m s⁻¹, is along AB, perpendicular to the river bank.

   (a) Sketch an appropriate triangle of velocities.

   (b) Find the value of v.

   (c) Find the value of θ.   [A]

9 A swimmer can move through still water at 1.2 m s⁻¹. She swims in a straight line in a river flowing at 0.8 m s⁻¹.

She travels from the point *A* to the point *B*, so that her resultant velocity makes an angle of 30° to the downstream bank, as shown in the diagram.

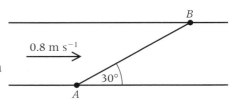

(a) Sketch an appropriate triangle of velocities.

(b) Find the magnitude of her resultant velocity. [A]

10 A model aircraft flies on a bearing of 010° with a speed of 50 km h⁻¹ relative to the air. A strong wind is blowing with a constant speed of *x* km h⁻¹. The resultant velocity of the aircraft is directed due north. The wind is blowing at an angle of 50° to the resultant velocity of the aircraft, as shown in the diagram.

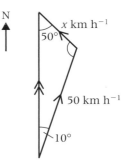

(a) Show that the value of *x* is approximately 11.3.

(b) **(i)** Find the northerly component of the 50 km h⁻¹ velocity.

(ii) Hence, or otherwise, find the magnitude of the resultant velocity of the aircraft. [A]

Key point summary

Formulae to learn:
Constant acceleration equations for two dimensions

$$\mathbf{r} = \mathbf{u}t + \frac{1}{2}\mathbf{a}t^2 \quad \text{or} \quad \mathbf{r} = \mathbf{u}t + \frac{1}{2}\mathbf{a}t^2 + \mathbf{r}_0$$

$$\mathbf{v} = \mathbf{u} + \mathbf{a}t$$

$$\mathbf{r} = \frac{1}{2}(\mathbf{u} + \mathbf{v})t \quad \text{or} \quad \mathbf{r} = \frac{1}{2}(\mathbf{u} + \mathbf{v})t + \mathbf{r}_0$$

1 A position vector enables paths to be plotted and interpreted. *p32*

2 Positions, velocities and accelerations can be expressed in the form $x\mathbf{i} + y\mathbf{j}$. *p36*

3 Magnitudes and directions of vectors can be found using trigonometric functions. *p36*

4 Constant acceleration equations can be used in two dimensions. *p41*

Test yourself	What to review

1 A ball moves so that its position, in metres, at time t seconds is given by

Section 3.2

$$\mathbf{r} = 5t\mathbf{i} + (6 + 9.5t - 5t^2)\mathbf{j}$$

where the unit vectors $\mathbf{i}$ and $\mathbf{j}$ are horizontal and vertical, respectively. The origin is at ground level.
   (a) Find the time when the ball hits the ground.
   (b) Plot the path of the ball.
   (c) Estimate the maximum height of the ball and the horizontal distance travelled by the ball from your plot.

2 A ship travels at 3 m s^{-1} on a bearing of 235°. Express this velocity in the form $a\mathbf{i} + b\mathbf{j}$, where $\mathbf{i}$ and $\mathbf{j}$ are unit vectors that are directed east and north, respectively.

Section 3.3

3 The velocity of an aeroplane is $(80\mathbf{i} + 50\mathbf{j})$ m s^{-1}, where $\mathbf{i}$ and $\mathbf{j}$ are unit vectors that are directed east and north, respectively. Find the speed of the aeroplane and the direction in which it is heading.

Section 3.3

4 The unit vectors $\mathbf{i}$ and $\mathbf{j}$ are perpendicular and lie in a horizontal plane. A particle moves from the origin. Its initial velocity was $(4\mathbf{i} + 6\mathbf{j})$ m s^{-1} and after 20 seconds its velocity is $(24\mathbf{i} + 46\mathbf{j})$ m s^{-1}. The acceleration of the particle is constant.
   (a) Find the acceleration of the particle.
   (b) Find the distance of the particle from the origin after accelerating for 30 seconds.

Section 3.4

5 An aeroplane is heading due north at 80 m s^{-1}. A wind blows due west at 25 m s^{-1}.
   (a) Find the resultant speed of the aeroplane.
   (b) Find the direction of the resultant velocity, giving your answer as a bearing.

Section 3.5

Test yourself ANSWERS

4 (a) $\mathbf{i} + 2\mathbf{j}$; **(b)** 1221 m.   **5 (a)** 83.8 m s^{-1}; **(b)** 342.6°.

2 $-3\cos 35°\mathbf{i} - 3\sin 35°\mathbf{j}$.   **3** 94.3 m s^{-1}, 058.0°.

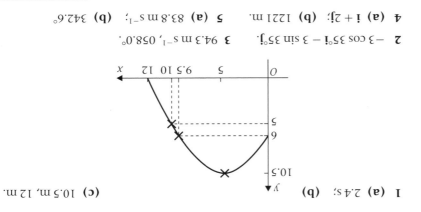

(c) 10.5 m, 12 m.   **1 (a)** 2.4 s; **(b)**

CHAPTER 4

Forces

Learning objectives

After studying this chapter you should be able to:
- identify the forces acting on a body
- draw force diagrams
- resolve forces into components
- find resultant forces
- write forces as vectors
- understand that the resultant force is zero when the forces are in equilibrium
- know that for equilibrium a body must be at rest or moving with a constant velocity
- use the friction inequality.

4.1 Introduction

In the earlier chapters of this book we have considered how to describe motion, using terms like velocity, acceleration and position. In this chapter we will consider forces. Forces cause motion or can act to keep objects at rest. An understanding of forces and how they cause motion is essential to be able to predict how an object in real life will move.

> Force is a vector quantity – it has magnitude and direction. The unit of force is the newton, which is abbreviated to N.

We will look at the different types of force that can act on a body, and learn how to add them together to find their sum, which is called their resultant. We will investigate a particle in equilibrium. This is where the resultant force on the particle is zero, so that the particle remains at rest or moves with a constant velocity. Finally we will be looking at frictional forces. We will be required to find unknowns such as forces, angles, and so on. In order to do this it is essential that we are able to draw a clear force diagram, which shows all the forces acting on the particle.

4.2 Types of force
Weight

Weight is a force, which is the effect of the earth's gravitational pull. If this force acted on its own on a body it would cause it to accelerate. This acceleration is approximately 9.8 m s^{-2} (denoted by g). The force, which causes this acceleration, is mg N, where m is the mass of the particle, in kilograms.

> The weight W of a body of mass m kg is given by
>
> $$W = mg$$
>
> where g is the acceleration due to gravity.
>
> On earth: $g = 9.8$ m s^{-2}.
> On the moon: $g = 1.6$ m s^{-2}.

Note that any two objects that are allowed to fall, will have the same acceleration and so fall at the same rate, even though the weight forces acting on them are different.

Force diagrams

As forces are vectors, we will represent them on diagrams by arrows, which show the direction of the force. We will indicate the size or magnitude of the force by writing this next to the arrow. The diagram shows a force of magnitude F N and the direction in which it acts. In a sketch the length of the arrow is not important, but if you were to solve problems by scale drawing, then the length of the vector or arrow would have to be proportional to the magnitude of the force.

Particle on a plane

Suppose a particle with a weight of W N is at rest on a smooth horizontal plane. As the particle does not fall, the weight W is being balanced by an equal force acting upwards. This force is called the normal reaction, R, and acts perpendicular to the plane.

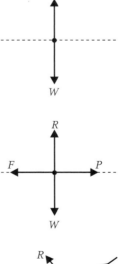

If we now place the particle on a rough horizontal plane, and pull the particle horizontally with a force of P N, then the roughness of the surface will oppose the motion. There will be a frictional force of F N, acting parallel to the plane, in the opposite direction. If the particle is at rest then $F = P$.

If we now consider the particle at rest on a rough plane, inclined at θ to the horizontal, then the frictional force will act up the slope, as shown in the diagram.

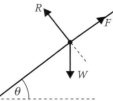

Strings and rods

A string will exert a force on an object if the string is taut, and this force, which is called a tension, will be directed along the string. For example, if a mass is suspended from a string, there will be a weight force acting downwards and a tension acting upwards.

Suppose a string is attached to a particle of weight W N, which rests on a smooth horizontal table. The only force, which the string can exert on the particle, will be in the same direction as the string itself. Furthermore, the string can only pull (and not push). Hence the tension, T, acts on the particle, as shown in the diagram.

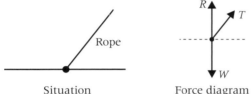

| Situation | Force diagram |

If a rod is attached to a particle instead of the string then the force of the rod on the particle could be in either of two directions, depending on whether it is pulling or pushing.

| Rod pulling | Rod pushing |

Here are some examples where the force diagram has been drawn to demonstrate particular situations.

1 A particle of weight W N sliding down a smooth plane inclined at α to the horizontal. Note that the term smooth means that there is no friction.

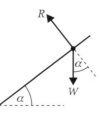

2 If the plane above was rough, instead of smooth, then a friction force would act parallel to the slope as shown in the diagram.

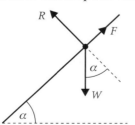

3 A particle of mass M kg at rest, suspended by a string.

4 A particle of mass M kg pulled across a rough horizontal plane by a string, inclined at 30° to the horizontal.

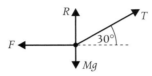

5 If the particle above was pulled on a smooth plane, there would be no friction force, as shown below.

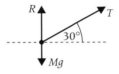

6 A and B are two points in the horizontal plane, a distance of 5 m apart. A particle, of weight W N, is attached by two strings, of length 3 m and 4 m, to the points A and B. The particle is at rest. The diagram shows the forces acting on the particle.

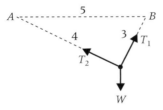

EXERCISE 4A

Draw force diagrams which show all the forces acting on the particle involved in each of the following situations:

1 A particle sliding up a rough plane inclined at α to the horizontal.

2 A particle of weight W N which is suspended from a fixed point by a string. The particle is held in equilibrium with the string at an angle θ to the vertical by a horizontal force P N.

3 A particle of mass m kg hangs in equilibrium supported by two light inextensible strings, inclined at 40° and 50° to the vertical.

4 A particle pushed up a rough plane, inclined at α to the horizontal, by a light rod, that is itself at an angle β to the slope.

5 A particle of weight W N is attached to one end of a light rod. The other end of the rod is fixed. A horizontal force P pulls the particle sideways, so that the rod makes an angle of 20° with the vertical.

4.3 Resultant forces

Adding two forces

Consider two forces, of magnitude F_1 and F_2, which act upon a particle. If we place these forces end to end, it can be seen that they have the same effect as a single force, of magnitude F. This force is known as the resultant force.

The resultant force will form the third side in a triangle of forces.

Worked example 4.1

Two forces of magnitudes 6 N and 5 N, act on a particle. The angle between the forces is 40°. Find the magnitude and direction of the resultant force.

Solution

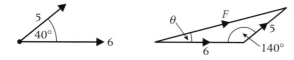

Using the cosine rule in the triangle of forces:

$$F^2 = 6^2 + 5^2 - 2 \times 6 \times 5 \times \cos 140°$$
$$F = 10.3 \text{ N}$$

Applying the sine rule:

$$\frac{\sin \theta}{5} = \frac{\sin 140}{F}$$
$$\theta = 18.1°$$

So the resultant force has magnitude 10.3 N and is at 18.1° to the 6 N force.

Worked example 4.2

Forces, of magnitude 14 N and 8 N, act on a particle. The resultant force has magnitude 17 N. Find the angle between the forces.

Solution

The required angle is θ, as shown. However, it is easier to first find the angle marked x, in the triangle of forces.

Using the cosine rule:

$$\cos x = \frac{14^2 + 8^2 - 17^2}{2 \times 14 \times 8}$$

$$x = 97.4°$$

The angle, θ, between the forces is:

$$180 - 97.4 = 82.6°$$

EXERCISE 4B

1 Find the magnitude of the resultant of the forces of magnitude F_1 and F_2, if θ is the angle between them. Also find the angle between the resultant and the force of magnitude F_1. Draw a diagram before starting each question.

   (a) $F_1 = 4$ N, $F_2 = 3$ N, $\theta = 90°$

   (b) $F_1 = 6$ N, $F_2 = 10$ N, $\theta = 60°$

   (c) $F_1 = 7$ N, $F_2 = 9$ N, $\theta = 75°$

   (d) $F_1 = 8$ N, $F_2 = 12$ N, $\theta = 170°$

   (e) $F_1 = 10$ N, $F_2 = 11$ N, $\theta = 160°$

2 Forces of magnitude F_1 and F_2 act on a particle. The resultant of the forces has magnitude R. Find the angle between the two forces acting on the particle in the following cases:

   (a) $F_1 = 60$ N, $F_2 = 80$ N, $R = 100$ N,

   (b) $F_1 = 11$ N, $F_2 = 17$ N, $R = 20$ N,

   (c) $F_1 = 7$ N, $F_2 = 10$ N, $R = 5$ N,

   (d) $F_1 = 8$ N, $F_2 = 7$ N, $R = 3$ N.

3 Forces of magnitude 6 N and 5 N act on a particle. What are the greatest and least values of the magnitude of the resultant of these forces?

4 The resultant of two forces of magnitude F_1 and F_2 has magnitude 20 N. If $F_1 = 8$ N and the angle between the two forces is 70°, find F_2.

5 The resultant of two forces of magnitude P and Q has magnitude 15 N. If $Q = 6$ N and the angle between the two forces is $60°$, find P.

6 The resultant of two forces of magnitude F_1 and F_2 has magnitude 3 N. If $F_1 = 7$ N and $F_2 = 5$ N, calculate the angle between the two forces.

7 The resultant of two forces of magnitude F_1 and F_2 has magnitude 12 N, and acts at an angle of $30°$ to F_1. If $F_1 = 6$ N find the magnitude of F_2 and direction of F_2.

4.4 Adding any number of forces

Three forces can be added by placing one force on the end of the other. The resultant of the forces is represented by the vector that can be drawn along the fourth side of the resulting quadrilateral, as shown below. The magnitude of the resultant, F, is given by the length of the fourth side.

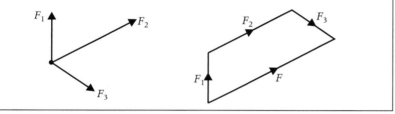

The quadrilateral can be divided into two triangles. The sine and cosine rules can then be used in the triangles to find the magnitude and direction of the resultant force.

Worked example 4.3

Find the magnitude of the resultant of the following set of forces:

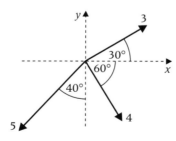

Solution

The diagram shows the three forces placed end to end to form a quadrilateral. The magnitude of the resultant force is F.

The angle between the 3 N force and the 4 N force is $90°$. Therefore Pythagoras' theorem can be used to find AC in triangle ABC.

$$AC^2 = 3^2 + 4^2,$$
$$AC = 5.$$

Also $\tan \theta = \frac{3}{4}$

$$\theta = 36.9°.$$
$$\alpha = 180 - \theta - 70 = 73.1°$$

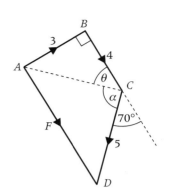

Using the cosine rule in triangle ACD gives:

$$F^2 = 5^2 + 5^2 - 2 \times 5 \times 5 \cos 73.1°$$
$$F = 5.96 \text{ N}$$

It is also possible to show that the angle between F and the x-axis is 76.6°.

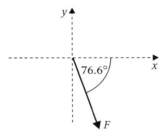

Using forces given in component form

The approach of Worked example 4.3 is a slow and complicated process in general. The resultant of a set of forces can be found more quickly when the forces are given in component form.

Forces can be expressed in the form $a\mathbf{i} + b\mathbf{j}$ in the same way that velocities and other vectors were expressed in chapter 3. For example a force that has magnitude 40 N and acts at 30° above the horizontal could be expressed as:

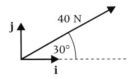

$$40 \cos 30°\mathbf{i} + 40 \sin 30°\mathbf{j}$$

First, we will concentrate on working with forces that are already expressed in vector form.

Worked example 4.4

Find the magnitude of the resultant of the following sets of forces:

$$\mathbf{F}_1 = (2\mathbf{i} + 3\mathbf{j}) \text{ N}, \mathbf{F}_2 = (\mathbf{i} - 2\mathbf{j}) \text{ N}, \mathbf{F}_3 = (2\mathbf{i} - 6\mathbf{j}) \text{ N}.$$

Solution

First find the resultant in terms of $\mathbf{i}$ and $\mathbf{j}$.

$$\mathbf{F} = \mathbf{F}_1 + \mathbf{F}_2 + \mathbf{F}_3$$
$$= 2\mathbf{i} + 3\mathbf{j} + \mathbf{i} - 2\mathbf{j} + 2\mathbf{i} - 6\mathbf{j}$$
$$= 5\mathbf{i} - 7\mathbf{j}$$

Then find the magnitude of the force.

$$F = \sqrt{5^2 + (-7)^2}$$
$$= \sqrt{74} = 8.60 \text{ N (to three significant figures)}$$

Worked example 4.5

The forces $(5\mathbf{i} + 12\mathbf{j})$ N and $(2\mathbf{i} + 4\mathbf{j})$ N act at a point.

(a) Find the magnitude of the resultant force.

(b) Find the angle between the resultant force and the unit vector $\mathbf{i}$.

Solution

(a) Resultant $= (5\mathbf{i} + 12\mathbf{j}) + (2\mathbf{i} + 4\mathbf{j})$
$$= 7\mathbf{i} + 16\mathbf{j}$$

Magnitude $= \sqrt{7^2 + 16^2}$
$$= 17.5 \text{ N to three significant figures}$$

(b) Angle, $\alpha = \tan^{-1}\left(\dfrac{16}{7}\right)$
$$= 66.4°$$

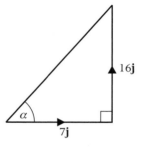

Note that column vectors can also be used for forces. The column vector $\begin{bmatrix} 4 \\ 7 \end{bmatrix}$ is equivalent to $4\mathbf{i} + 7\mathbf{j}$.

EXERCISE 4C

1 Find the magnitude of the resultant of the following sets of forces, by forming a quadrilateral of forces.

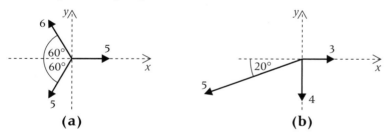

(a)   **(b)**

2 Find the magnitude of the resultant of the forces $(2\mathbf{i} + \mathbf{j})$ N, $(3\mathbf{i} - 2\mathbf{j})$ N and $(-2\mathbf{i} + 4\mathbf{j})$ N.

3 The resultant of the forces $(2\mathbf{i} + \mathbf{j})$ N, $3\mathbf{j}$ N, $(2\mathbf{i} + 4\mathbf{j})$ N, $(6\mathbf{i} + b\mathbf{j})$ N and $(a\mathbf{i} + \mathbf{j})$ N is $(3\mathbf{i} + 4\mathbf{j})$ N, where $\mathbf{i}$ and $\mathbf{j}$ are perpendicular unit vectors. Find a and b.

4 Three forces $(3\mathbf{i} + 5\mathbf{j})$ N, $(4\mathbf{i} + 11\mathbf{j})$ N, $(2\mathbf{i} + \mathbf{j})$ N act at a point. Given that $\mathbf{i}$ and $\mathbf{j}$ are perpendicular unit vectors find:

(a) the resultant of the forces in the form $a\mathbf{i} + b\mathbf{j}$,

(b) the magnitude of this resultant,

(c) the angle that the resultant makes with the unit vector $\mathbf{i}$.   [A]

5 Two forces $(3\mathbf{i} + 2\mathbf{j})$ N and $(-5\mathbf{i} + \mathbf{j})$ N act at a point. Find the magnitude of the resultant of these forces and determine the angle which the resultant makes with the unit vector $\mathbf{i}$.   [A]

6 Two forces $\begin{bmatrix} 7 \\ 1 \end{bmatrix}$ and $\begin{bmatrix} -6 \\ 3 \end{bmatrix}$ act at a point. Calculate the magnitude of the resultant of these two forces.

7 The forces $\mathbf{F}_1 = \begin{bmatrix} a \\ -2a \end{bmatrix}$ and $\mathbf{F}_2 = \begin{bmatrix} 2 \\ 4 \end{bmatrix}$ act on a particle. The resultant of $\mathbf{F}_1$ and $\mathbf{F}_2$ has magnitude 4 N. Find the possible values of a.

8 Three forces $(\mathbf{i} + \mathbf{j})$ N, $(-5\mathbf{i} + 3\mathbf{j})$ N and $\lambda\mathbf{i}$ N, where $\mathbf{i}$ and $\mathbf{j}$ are perpendicular unit vectors, act at a point. Express the resultant in the form $(a\mathbf{i} + b\mathbf{j})$ and find its magnitude in terms of λ. Given that the resultant has magnitude 5 N, find the two possible values of λ.

Take the larger value of λ and find the tangent of the angle between the resultant and the unit vector $\mathbf{i}$.   [A]

4.5 Resolving forces

A force can be divided into two mutually perpendicular components whose vector sum is equal to the given force.

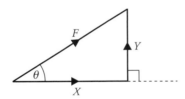

From the right-angled triangle:

$X = F \cos \theta$

and

$Y = F \sin \theta$

This process enables us to write forces in the form $a\mathbf{i} + b\mathbf{j}$. In the case above we would write:

$F \cos \theta\mathbf{i} + F \sin \theta\mathbf{j}$

Worked example 4.6

A force of 10 N acts at 60° below the horizontal. The unit vectors $\mathbf{i}$ and $\mathbf{j}$ are horizontal and vertical, respectively. Write the force in terms of the unit vectors $\mathbf{i}$, and $\mathbf{j}$.

Solution

The diagram shows the force and the unit vectors.

The horizontal component of the force is 10 cos 60°.

The vertical component of the force is −10 sin 60°

Hence the force can be written as:

10 cos 60°**i** − 10 sin 60°**j** = 5**i** − 8.66**j** (to three significant figures)

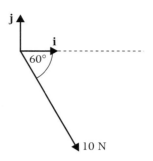

This can also be expressed using a column vector:

$$\begin{bmatrix} 10 \cos 60° \\ -10 \sin 60° \end{bmatrix} = \begin{bmatrix} 5 \\ -8.66 \end{bmatrix}$$

Worked example 4.7

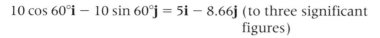

Find the magnitude and direction of the resultant of the set of forces given below.

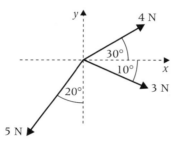

Solution

Resolving horizontally:

4 cos 30° + 3 cos 10° − 5 sin 20° = 4.708

Resolving vertically:

4 sin 30° − 3 sin 10° − 5 cos 20° = −3.219

The resultant force has magnitude $F = \sqrt{4.708^2 + 3.219^2} = 5.70$ N (to three significant figures) and acts at an angle

$\tan^{-1}\left(\dfrac{3.219}{4.708}\right) = 34.4°$ below the positive *x*-axis.

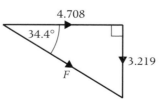

It is a good idea to draw a diagram in such cases, as it makes your intention clear

EXERCISE 4D

1 Find the components of the following forces in the directions of the *x*-axis and *y*-axis.

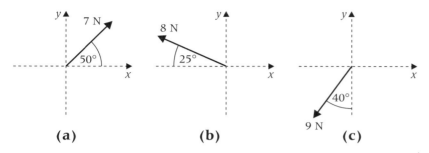

(a)   **(b)**   **(c)**

2 Express the forces in question 1 in the form $a\mathbf{i} + b\mathbf{j}$, if $\mathbf{i}$ is directed along the *x*-axis and $\mathbf{j}$ is directed along the *y*-axis.

3 Express each force in question 1 as a column vector.

4 Find the components of the following forces in the directions of the *x*-axis and *y*-axis.

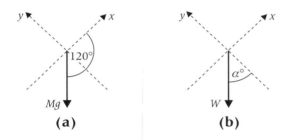

(a)   **(b)**

5 Find the magnitude and the direction of the resultant force in each of the following cases:

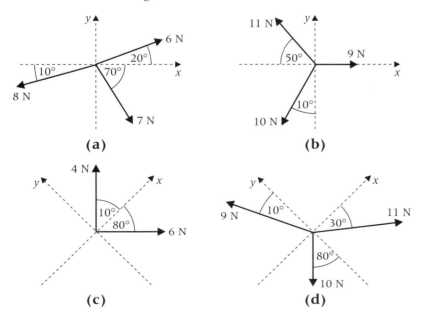

(a)   **(b)**

(c)   **(d)**

6 Three horizontal forces each of magnitude 10 N act in the directions of the bearings 040°, 160° and 280°. Find the magnitude of their resultant.

7 The following diagrams show a particle on an inclined plane. Find the components of the weight of the particle in the directions of the *x*-axis and *y*-axis.

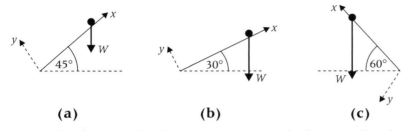

(a)   **(b)**   **(c)**

8 A particle of mass *m* kg lies at rest on a rough plane, inclined at α to the horizontal. What is the component of the weight down the plane?

4.6 Equilibrium

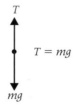

> If a set of forces act on a particle such that their resultant is zero, then the particle is said to be in equilibrium: that is, no unbalanced forces act on the particle.

When a particle is in equilibrium it will either remain at rest or move with a constant velocity.

Two forces

If only two forces act on a particle, which is in equilibrium, then the forces must be equal and opposite. For example, a particle of mass *m* kg rests in equilibrium, suspended by a light inextensible string.

$$T = mg$$

Three forces

If forces, of magnitude F_1, F_2 and F_2, have a zero resultant, then when the forces are placed end to end they must form a triangle.

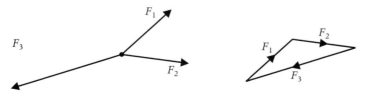

The sine rule and the cosine rule can be used in this triangle.

Worked example 4.8

A particle of mass 10 kg hangs in equilibrium suspended by two light inextensible strings, as shown. Find the tensions in the strings.

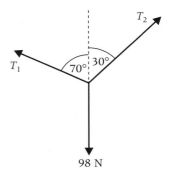

Solution

The diagram below shows how the forces must form a triangle, as they are in equilibrium.

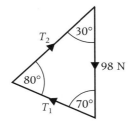

From the sine rule:

$$\frac{T_1}{\sin 30°} = \frac{T_2}{\sin 70°} = \frac{98}{\sin 80°}$$

$$T_1 = \frac{98 \sin 30°}{\sin 80°} \quad \text{and} \quad T_2 = \frac{98 \sin 70°}{\sin 80°}$$

$T_1 = 49.8$ N, $T_2 = 93.5$ N (to three significant figures)

Worked example 4.9

The set of forces shown below is in equilibrium. Find P and θ.

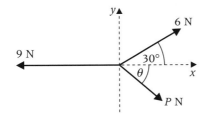

Solution

As the forces are in equilibrium, they can be arranged to form a triangle as shown in the diagram.

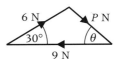

Using the cosine rule in this triangle gives:

$$P^2 = 6^2 + 9^2 - 2 \times 6 \times 9 \times \cos 30°$$

$$P = 4.84 \text{ N}$$

From the sine rule:

$$\frac{\sin \theta}{6} = \frac{\sin 30°}{4.84}$$

$$\theta = 38.3°$$

4

More than three forces

The previous two problems were solved using the sine and cosine rules only. Alternative solutions can be produced by resolving the forces in two perpendicular directions, for example horizontal and vertical. This method is particularly useful when we have more than three forces acting on the particle.

Worked example 4.10

The forces in the diagram are in equilibrium. Find P and the angle θ.

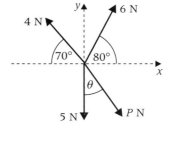

Solution

Resolving vertically:

$$P \cos \theta + 5 = 4 \sin 70° + 6 \sin 80° \qquad [1]$$
$$P \cos \theta = 4.668$$

Resolving horizontally:

$$P \sin \theta + 6 \cos 80° = 4 \cos 70° \qquad [2]$$
$$P \sin \theta = 0.326$$

Simultaneous equations like [1] and [2] occur frequently in mechanics. They can be solved using the trigonometric identities

$$\sin^2 \theta + \cos^2 \theta \equiv 1 \qquad \tan \theta \equiv \frac{\sin \theta}{\cos \theta}$$

Hence $\tan \theta = \dfrac{0.326}{4.668}$

$$\theta = 4.0°$$

and $P^2 = 4.668^2 + 0.326^2$

$$P = 4.68 \text{ N}$$

Worked example 4.11

A particle of mass 5 kg is at rest on a rough plane inclined at 50° to the horizontal. Find the normal reaction and the frictional force on the particle.

Solution

The diagram shows the forces acting on the particle.

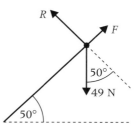

Resolving parallel to the plane:

$$F = 49 \sin 50° = 37.5 \text{ N}$$

Resolving perpendicular to the plane:

$$R = 49 \cos 50° = 31.5 \text{ N}$$

The choice of whether to solve a problem by resolving or by drawing the triangle of forces and using trigonometry depends upon the particular situation involved. In the preceding examples, two demonstrated the use of trigonometry, and two showed how to find the solution by resolving. Whereas some problems are much easier done one way than the other, either technique will work equally well for certain types of problem. Worked example 4.11, for instance could be solved equally well by either method.

EXERCISE 4E

1 Each of the following sets of forces is in equilibrium. Find the magnitudes F_1 and F_2.

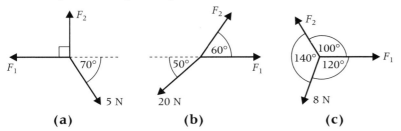

(a)   (b)   (c)

2 Each of the following sets of forces is in equilibrium. Find F and the angle θ.

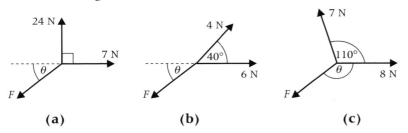

(a)   (b)   (c)

3 Each of the following sets of forces is in equilibrium. Find F_1 and F_2.

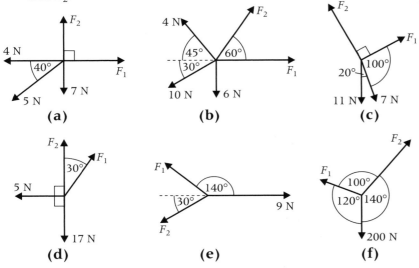

(a)   (b)   (c)

(d)   (e)   (f)

4 A particle is in equilibrium, subject to forces $(6\mathbf{i} + \mathbf{j})$ N, $(2\mathbf{i} + 3\mathbf{j})$ N and **P**.

   (a) Find **P** in terms of **i** and **j**.

   (b) Find the magnitude of **P** and the angle between **P** and **i**.

5 The set of forces $(a\mathbf{i} + 2\mathbf{j})$ N, $(6\mathbf{i} - 3\mathbf{j})$ N, $(4\mathbf{i} + b\mathbf{j})$ N, and $(\mathbf{i} + \mathbf{j})$ N is in equilibrium. Find a and b.

6 Forces $\mathbf{F}_1$, $\mathbf{F}_2$ and $\mathbf{F}_3$ are in equilibrium.

   (a) If $\mathbf{F}_1 = (\mathbf{i} + \mathbf{j})$ N, and $\mathbf{F}_3 = (7\mathbf{i} + 8\mathbf{j})$ N, find $\mathbf{F}_2$.

   (b) If $\mathbf{F}_1 = (2\mathbf{i} + \mathbf{j})$ N, $\mathbf{F}_2 = (-4\mathbf{i} + \mathbf{j})$ N, find $\mathbf{F}_3$.

7 The forces $\mathbf{F}_1 = \begin{bmatrix} x \\ 3 \end{bmatrix}$, $\mathbf{F}_2 = \begin{bmatrix} -7 \\ y \end{bmatrix}$ and $\mathbf{F}_3 = \begin{bmatrix} 6 \\ y \end{bmatrix}$ are in equilibrium. Find x and y.

8 Three forces $\mathbf{F}_1$, $\mathbf{F}_2$ and $\mathbf{F}_3$ are in equilibrium. If $\mathbf{F}_1 = \begin{bmatrix} 4 \\ -7 \end{bmatrix}$ and $\mathbf{F}_2 = \begin{bmatrix} -6 \\ 8 \end{bmatrix}$, find the magnitude of $\mathbf{F}_3$.

9 Horizontal forces each of magnitude 10 N act in the direction of the bearings 040°, 160° and 280°. Are these forces in equilibrium?

10 A mass of 5 kg is suspended by two light, inextensible strings. The angles between the strings and the vertical are 30° and 60°. Find the tensions in the strings.

11 A mass of 10 kg is suspended by two light, inextensible strings. The angles between the strings and the vertical are 40° and 20°. Find the tensions in the strings.

12 A particle of weight 10 N is suspended by two light, inextensible strings. The tension in one string is 5 N, which acts at 20° to the vertical. Find the tension in the second string and the angle between it and the vertical.

13 A particle of weight 10 N is in equilibrium on a smooth plane, inclined 40° to the horizontal. A horizontal force P acts on the particle. Find P and the normal reaction between the plane and the particle.

14 A particle of weight 10 N is suspended from a fixed point by a light inextensible string. A horizontal force of 5 N also acts on the particle. Find the tension in the string and the angle between the string and the vertical.

15 Three forces act upon a particle, which is in equilibrium. If the magnitudes of the forces are 4 N, 5 N, and 6 N, find the angles between the forces.

16 A load of mass 50 kg is supported, in equilibrium, by two ropes. One is at an angle of 30° to the vertical and the other is horizontal, as shown in the diagram. The tensions in these ropes are T_1 newtons and T_2 newtons, respectively.

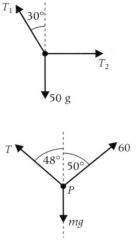

(a) Show that $T_1 = 566$ N, correct to three significant figures.

(b) Find T_2.   [A]

17 A particle, P, of mass m kg, is held in equilibrium by two strings. One string is inclined at 50° to the vertical and exerts a force of 60 newtons on the particle. The other string exerts a force of magnitude T newtons at an angle of 48° to the vertical. The forces that act on the particle are shown in the diagram opposite.

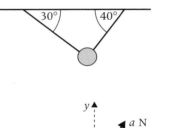

4

(a) Find T.

(b) Find m.   [A]

18 An object of mass 3 kg is suspended by two light, inextensible strings. The strings make angles of 30° and 40° and the horizontal, as shown in the diagram.

Find the magnitude of the tension in each string.

19 A particle is at a point O on a smooth horizontal surface. It is acted on by three horizontal forces of magnitudes 6 N, 8 N and a N. Relative to horizontal axes Ox and Oy, the directions of these three forces are shown in the diagram. The resultant, R, of these forces acts along the line Oy.

(a) Show that $a = 4$.

(b) Find the magnitude of R.   [A]

20 A particle P, lies on a smooth horizontal surface. It is acted on by two horizontal forces of magnitudes 25 N and 20 N. Relative to horizontal axes Px and Py, the directions of these two forces are as shown in the diagram. A third horizontal force **F** is required to keep P in equilibrium.

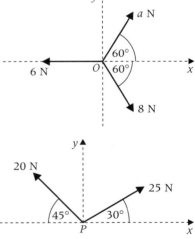

(a) Express the force of magnitude 25 N as a column vector, giving its components to one decimal place.

(b) Obtain F as a column vector, giving its components to one decimal place.   [A]

21 Four boys are playing a 'tug of war' game, each pulling horizontally on a rope attached to a light ring. Boy A pulls with a force of $(92\mathbf{i} - 33\mathbf{j})$ N, boy B with force $(66\mathbf{i} + 62\mathbf{j})$ N and by C with force $(-70\mathbf{i} + 99\mathbf{j})$ N, where $\mathbf{i}$ and $\mathbf{j}$ are perpendicular unit vectors. Given that the ring is in equilibrium, find the force exerted by boy D, and its magnitude.

4.7 Friction

Consider a body of mass m, placed on a rough horizontal plane. Suppose a horizontal force of magnitude P is applied. If the body remains in equilibrium, then the frictional force, F, will be such that $F = P$.

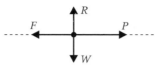

If P is steadily increased and the body remains at rest, then F increases also, so that $F = P$ remains true. However F can only increase up to a certain limit, F_{max}.

> The force between the surfaces in contact and the type of surface determine how large the frictional force can become.

Racing cars make use of this principle. The downwards force on the vehicle is increased by incorporating an aerofoil into the design. This improves the grip of the tyres on the road surface, so that drivers can take corners at greater speed without skidding.

The vertical force between the surfaces is the normal reaction R. If a weight is placed on top of a body so that the normal reaction is doubled, it will be found that F_{max} will also double. That is, F_{max} is proportional to R (the normal reaction).

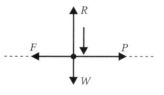

> $F_{max} = \mu R$
>
> where μ is a constant, which depends only on the roughness of the surface, known as the coefficient of friction.

If P is increased even further then slipping will occur. The frictional force cannot increase further and experiments show that the frictional force remains constant at its maximum throughout the motion.

N.B.   (i)   When the frictional force does equal its maximum it is often said to be **limiting**.

      (ii)   The special case $\mu = 0$ means $F_{max} = 0$, which corresponds to a smooth plane.

Horizontal planes

Worked example 4.12

A particle of mass 10 kg is placed on a rough horizontal plane. A horizontal force P acts on the particle. P is increased until the particle is on the point of sliding, which occurs when $P = 10$ N. Find the coefficient of friction between the particle and the plane.

Solution

The diagram shows the forces acting on the particle.

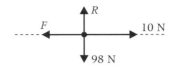

Resolving vertically: $R = 98$ N

Resolving horizontally: $F = 10$ N

If motion is just about to occur, friction is limiting, $F = F_{\text{max}}$
so $F = \mu R$.

$$10 = 98\mu$$
$$\mu = \frac{10}{98} = \frac{5}{49}$$

Note that μ does not have units.

Worked example 4.13

In the following situations a body of mass 10 kg is placed on a rough horizontal plane. If $\mu = \frac{1}{2}$ in each case determine whether motion will occur.

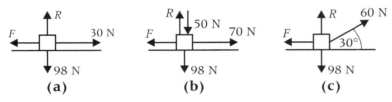

(a)   (b)   (c)

Solution

(a) Resolving vertically: $R = 98$ N

$$F_{\text{max}} = 0.5 \times 98 = 49 \text{ N}$$

For equilibrium $F = 30 < 49$, so no motion will occur.

(b) Resolving vertically: $R = 98 + 50 = 148$ N

$$F_{\text{max}} = 0.5 \times 148 = 74 \text{ N}$$

For equilibrium $F = 70 < 74$, so again no motion will occur.

(c) Resolving vertically: $R + 60 \sin 30° = 98$

$$R = 68 \text{ N}$$
$$F_{\text{max}} = 0.5 \times 68 = 34 \text{ N}$$

For equilibrium $F = 60 \cos 30° = 52.0 > 34$, which is not possible, so sliding will occur.

Use of the inequality $F \leqslant \mu R$

In all possible cases of a particle in equilibrium on a rough plane, the frictional force has a limiting value so the following inequality is always true:

$$F \leqslant \mu R$$

This inequality itself can be used to good effect.

Worked example 4.14

A particle, of mass 10 kg, is at rest on a rough horizontal plane. A force, of magnitude of P N, is applied in the direction shown. If $\mu = \frac{1}{2}$, what is the greatest possible value of the magnitude of P such that motion does not occur?

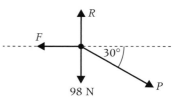

Solution

Resolving horizontally: $F = P \cos 30°$.

Resolving vertically: $R = 98 + P \sin 30°$.

If we substitute for F and R in $F \leq \mu R$

$$P \cos 30° \leq \tfrac{1}{2}(98 + P \sin 30°)$$
$$P \cos 30° \leq 49 + \tfrac{1}{2}P \sin 30°$$
$$P \cos 30° - \tfrac{1}{2}P \sin 30° \leq 49$$
$$P(\cos 30° - \tfrac{1}{2}\sin 30°) \leq 49$$
$$P \leq \frac{49}{\cos 30° - \tfrac{1}{2}\sin 30°}$$
$$P \leq 79.5 \text{ N}$$

So the greatest value of P is 79.5 N (to 3 significant figures).

Worked example 4.15

A particle of weight W is at rest on a rough horizontal surface. A force of magnitude $\frac{1}{2}W$ is applied to the particle as shown. Find the coefficient of friction, μ, between the particle and the surface, if the particle is on the point of sliding.

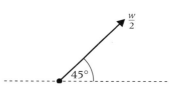

Solution

The diagram shows the forces acting on the particle.

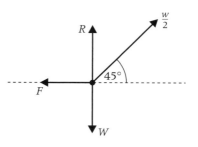

Resolving horizontally: $F = \dfrac{W}{2} \cos 45°$

Resolving vertically: $R + \dfrac{W}{2} \sin 45° = W$

$$R = W - \frac{W}{2} \sin 45°$$
$$= W(1 - \tfrac{1}{2} \sin 45°)$$

Substituting for F and R in $F = \mu R$ we get:

$$\frac{W}{2} \cos 45° = \mu W(1 - \tfrac{1}{2} \sin 45°)$$

$$\mu = \frac{\tfrac{1}{2} \cos 45°}{1 - \tfrac{1}{2} \sin 45°} = 0.547$$

Note that in this case $F = \mu R$ can be used because the particle is on the point of sliding.

EXERCISE 4F

1 A horizontal force of magnitude 5 N is applied to a body of mass 20 kg which is at rest on a rough horizontal plane. Find the coefficient of friction, given that friction is limiting in this position.

2 In the following situations a particle of mass 4 kg is placed on a rough horizontal plane. If $\mu = 5/7$ determine whether motion will occur.

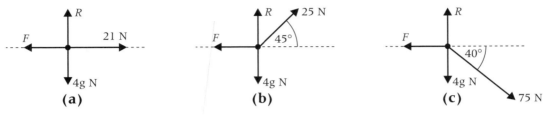

(a)　　　　　　**(b)**　　　　　　**(c)**

3 A particle of weight 10 N is at rest on a rough horizontal plane. The particle is pulled by a light, inextensible string, inclined at an angle of 20° to the plane. If the tension in the string is 5 N, find the least value of the coefficient of friction correct to three significant figures.

4 The diagram shows a particle of mass of 6 kg at rest on a rough horizontal plane, subject to an external force of magnitude P N. What is the greatest value of P, if $\mu = 2/3$?

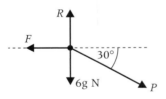

5 Horizontal forces, each of magnitude 100 N act in the direction of the bearings 050°, 170°, and 260°. Find the magnitude of the resultant of these forces. If these forces act on a particle, of mass 10 kg, which is in equilibrium on a rough horizontal plane, find the magnitude of frictional force, which acts on the particle. Find also the least value of the coefficient of friction, giving your answer correct to two significant figures.

6 The coefficient of friction between a sledge and a snowy surface is 0.2. The combined mass of a child and the sledge is 45 kg. What is the least horizontal force necessary to pull the sledge along the horizontal surface at a constant speed?

7 A sledge of mass 12 kg is on level ground.
　(a) A horizontal force of 10 N will keep the sledge moving at a constant speed. Find the value of the coefficient of friction.
　(b) A girl of mass 25 kg sits on the sledge. Find the least horizontal force required to keep the sledge moving at a constant speed.

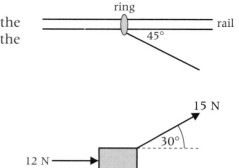

8 A small ring, of weight w N, is threaded on a horizontal curtain rail. A light, inextensible string is pulling it along the rail. The tension in the string is equal to $2w$ N. Show that the least value of the coefficient of friction is 0.586, correct to three decimal places.

9 The diagram shows a small box resting on a rough horizontal surface.

The box is of weight W newtons. It is pushed with a horizontal force of 12 newtons, and pulled with a force of 15 newtons at an angle of 30° to the horizontal.

The box rests in limiting equilibrium.

(a) Draw a diagram to show all the forces acting on the box.

(b) Show that the frictional force acting on the box is approximately 25 newtons.

(c) The coefficient of friction between the box and the surface is $\frac{1}{3}$.
Find the normal reaction force between the box and the surface.

(d) Find the value of W. [A]

10

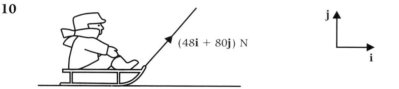

The diagram shows a child on a sledge which is being pulled at **constant velocity** across a snowy horizontal field. The tension in the rope pulling the sledge is $(48\mathbf{i} + 80\mathbf{j})$ N, where the unit vectors $\mathbf{i}$ and $\mathbf{j}$ are horizontal and vertical, respectively, as shown in the diagram.

(a) Draw a diagram showing all the forces which act on the sledge.

(b) State the magnitude of the frictional force acting on the sledge.

(c) The coefficient of friction between the sledge and the surface is 0.4. Show that the normal reaction force between the sledge and the field is of magnitude 120 N.

(d) Find the total weight of the child and the sledge. [A]

11 The coefficient of friction between a particle, of weight w, and a horizontal plane is μ. The particle is in equilibrium subject to a force of magnitude P N which acts at an angle of θ below the horizontal. Show that:

$$P \leqslant \frac{\mu w}{\cos\theta - \mu\sin\theta}$$

4.8 Friction on inclined planes

Consider a particle, of mass m kg, which is at rest on a rough plane inclined at an angle α to the horizontal.

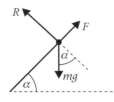

> The diagram shows the forces acting on the particle. The friction force must act up the slope to maintain equilibrium.

Resolving parallel to the plane: $\qquad F = mg \sin \alpha$

Resolving perpendicular to the plane: $R = mg \cos \alpha$

However

$$F \leqslant \mu R$$

so

$$mg \sin \alpha \leqslant \mu\, mg \cos \alpha$$

hence:

$$\frac{\sin \alpha}{\cos \alpha} \leqslant \mu$$
$$\tan \alpha \leqslant \mu$$

This result tells us that the particle will slip if the angle becomes too steep. The particle will be on the point of slipping when

$$\tan \alpha = \mu$$

or

$$\alpha = \tan^{-1} \mu$$

Worked example 4.16

A block of mass 3 kg is placed on a rough horizontal table. The table is gradually tilted until the particle begins to slip. The block is on the point of slipping when the table is inclined at an angle of 41° to the horizontal. Find the coefficient of friction.

Solution

The diagram shows the forces acting on the block.

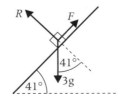

Resolving parallel to the plane: $\qquad F = 3\,g \sin 41° = 19.3$ N

Resolving perpendicular to the plane $\quad R = 3\,g \cos 41° = 22.2$ N

But if the particle is on the point of sliding then friction is limiting so

$$F = \mu R$$
$$19.3 = 22.2\mu$$
$$\mu = 0.869 \text{ (to three significant figures)}$$

Note: We could have used $\mu = \tan \alpha$

Suppose this table is now placed at an angle of 50° to the horizontal. The block would not be able to rest in equilibrium unaided. So, suppose a force, of magnitude P, is applied to the block so that it acts up the plane. We will now find the least value of P to maintain equilibrium.

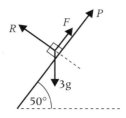

Resolving parallel to plane: $\qquad F + P = 3\,g \sin 50°$

$$F = 3\,g \sin 50° - P$$

Resolving perpendicular to plane: $\quad R = 3\,g \cos 50°$

Here we can use $F \leqslant \mu R$, and substitute for F and R.

$$3\,g \sin 50° - P \leqslant \mu\, 3\,g \cos 50°$$

$$P \geqslant 3\,g \sin 50° - \mu\, 3\,g \cos 50°$$

$$P \geqslant 6.09 \text{ N (to three significant figures)}$$

So the **minimum** value of P is 6.09 N if the particle is to remain at rest.

If the force P is too large, however, the particle will be pulled up the plane so we will now find the **greatest** value of P such that the particle will remain in equilibrium.

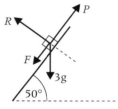

The method used to solve this problem will be exactly the same as that used to find the minimum value of P. There is one fundamental difference, however, which can be seen in the diagram; the frictional force now acts down the slope. If we are considering pulling the particle up the slope, then friction will oppose the motion and so act down the slope.

Resolving parallel to plane: $\qquad P = F + 3\,g \sin 50°$

Resolving perpendicular to plane: $\quad R = 3\,g \cos 50°$

Substituting into $F \leqslant \mu R$

$$P - 3\,g \sin 50° \leqslant \mu\, 3\,g \cos 50°$$

$$P \leqslant 3\,g \sin 50° + \mu\, 3\,g \cos 50°$$

$$P \leqslant 38.9 \text{ N}$$

Worked example 4.17

A particle of mass 10 kg is at rest on a rough plane inclined at 30° the horizontal. A horizontal force of magnitude 10 N acts on the particle.

(a) Find the magnitude of the friction force on the particle.

(b) The coefficient of friction between the particle and the slope is μ. Find an inequality that μ must satisfy.

Solution

(a) The diagram shows the forces acting on the particle.
Resolving parallel to the slope:

$$F + 10 \cos 30° = 98 \sin 30°$$
$$F = 98 \sin 30° - 10 \cos 30°$$
$$= 40.3 \text{ N (to 3 significant figures)}$$

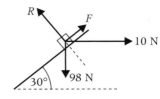

(b) Resolving perpendicular to the slope gives:

$$R = 98 \cos 30° + 10 \sin 30°$$
$$= 89.9 \text{ N} \quad \text{(to 3 significant figures)}$$

The friction inequality states that

$$F \leqslant \mu R$$

or

$$\mu \geqslant \frac{F}{R}.$$

In this case,

$$\mu \geqslant \frac{98 \sin 30° - 10 \cos 30°}{98 \cos 30° + 10 \sin 30°}$$

$$\mu \geqslant 0.449 \text{ (to 3 significant figures)}$$

EXERCISE 4G

1 In the following situations a particle of mass m kg is placed on a rough plane inclined at an angle α to the horizontal. Determine whether the particle will slide.

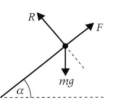

(a) $\alpha = 20°$, $\mu = 0.1$

(b) $\alpha = 30°$, $\mu = 0.75$

(c) $\alpha = 50°$, $\mu = 0.9$

2 The situations below show a particle of mass 10 kg at rest on an inclined plane, where the coefficient of friction is 0.1. Find the magnitude, P, of the applied force, if the particle is on the point of sliding down the plane.

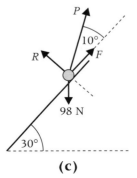

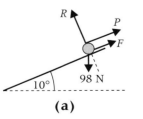

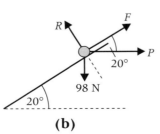

(a)

(b)

(c)

3 The situations below show a particle of mass 10 kg at rest on a rough inclined plane, where the coefficient of friction is 0.1. Find the magnitude, P, of the force applied to the particle, if it is on the point of moving up the plane.

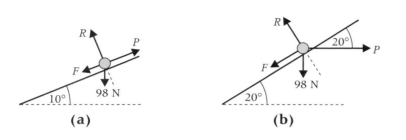 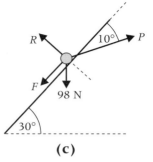

(a)   (b)   (c)

4 A particle of mass 10 kg is placed on a rough plane, inclined at 45° to the horizontal. If $\mu = 0.5$ find the least force required to keep the particle in equilibrium, if the force acts upwards, along the line of greatest slope.

5 A particle of mass 6 kg is at rest on a rough plane of inclination α to the horizontal. Find the greatest horizontal force that can be applied to the particle if it is to remain in equilibrium, in each of the following cases:
   (a) $\alpha = 20°$, $\mu = 0.1$
   (b) $\alpha = 30°$, $\mu = 0.9$
   (c) $\alpha = 50°$, $\mu = 0.7$.

6 A particle, of mass 12 kg, is at rest on a rough plane, inclined at an angle α to the horizontal. Find the least force, which is parallel to the plane, that must be applied to the particle, if it is to remain in equilibrium, in each of the following cases:
   (a) $\alpha = 20°$, $\mu = 0.1$
   (b) $\alpha = 30°$, $\mu = 0.5$
   (c) $\alpha = 50°$, $\mu = 1.0$.

7 A particle of mass 8 kg is at rest on a rough plane inclined at an angle of 30° to the horizontal. A horizontal force of 20 N acts on the particle as shown. Find the magnitude of the friction force and the normal reaction on the particle. What is the least value of μ?

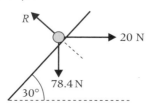

8 The situations below show a particle of mass 4 kg at rest on an inclined plane, subject to a given external force. Draw a force diagram showing all the forces acting on the particle. Find the normal reaction, the friction force, and the range of values of the coefficient of friction in each case.

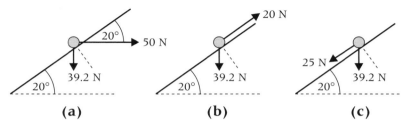

(a)   **(b)**   **(c)**

4

Key point summary

Formula to learn:

$F \leqslant \mu R$

1 Force is a vector quantity.	*p54*
2 The resultant force is the sum of two or more forces. It forms a triangle of forces with two forces and a quadrilateral with three forces.	*p58*
3 Resultant forces can often be found more quickly using forces in component form.	*p61*
4 A force can be divided into two mutually perpendicular components whose vector sum is equal to the given force.	*p63*
5 A particle is in equilibrium if the resultant force acting on it is zero.	*p66*
6 The frictional force between two surfaces in contact will act in a direction to oppose relative motion.	*p72*
7 The magnitude of the frictional force will never be greater than what is necessary to prevent motion.	*p72*
8 The maximum value of the frictional force is given by: $F_{max} = \mu R$.	*p72*
9 If motion takes place then the magnitude of friction force will be μR.	*p72*

Test yourself	**What to review**

1 Draw diagrams to show the forces acting on each of the following. You should include air resistance where appropriate, and model each body as a particle.

   (a) A golf ball at its maximum height.

   (b) A cyclist travelling up a slope at a constant speed.

   (c) A child on a swing at her lowest position.

Section 4.2

2 The diagram shows three forces and the perpendicular unit vectors **i** and **j**.

   (a) Find the resultant of these three forces in terms of the unit vectors **i** and **j**.

   (b) Find the magnitude of the resultant of these three forces and draw a diagram to show the direction in which it acts.

   (c) When a fourth force acts at the same point the forces are in equilibrium. Find the magnitude of this force and describe the direction in which it acts.

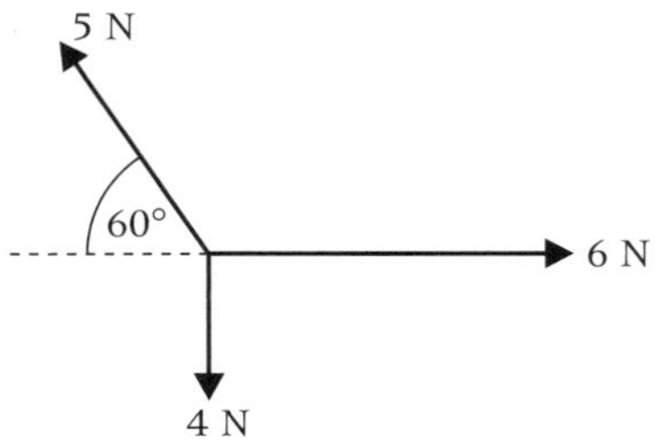

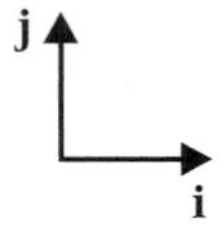

Section 4.3 and 4.4

3 The diagram shows an object of mass 50 kg, which is supported by two cables. Find the tension in each of the supporting cables.

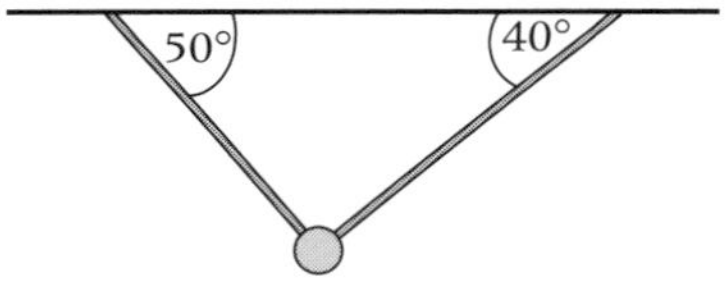

Section 4.5

4 The diagram shows a spring fixed to a wall and a block of mass 20 kg, that is at rest on a rough horizontal plane. A string attached to the block passes over a small smooth pulley and is attached to a second block of mass 5 kg. The situation is shown in the diagram.

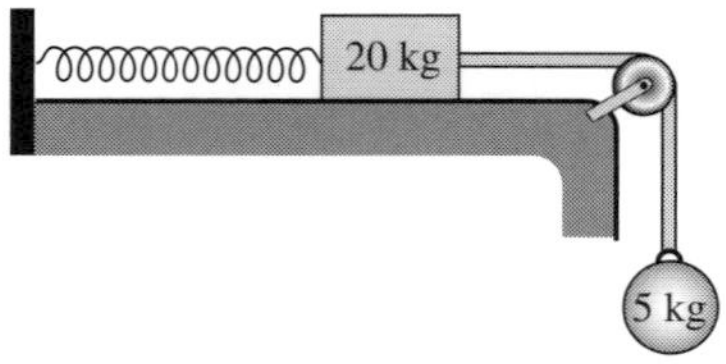

The tension in the spring is 8 N when the block is on the point of sliding towards the pulley.

   (a) Find the coefficient of friction between the block and the plane.

   (b) Describe what happens to the 20 kg block if the string attached to it is cut. Give reasons to support your answer.

Section 4.7

5 A particle, of mass 5 kg, is at rest on a slope inclined at an angle of 48° to the horizontal. A force, of magnitude 10 N, that is directed up the slope acts on the particle.

   (a) Find the magnitude of the friction force acting on the particle.

   (b) Find an inequality that the coefficient of friction between the particle and the slope must satisfy.

Section 4.7

1 (a) Air resistance

mg

(b) R Forward force

Air resistance

mg

(c) T

Air resistance

mg

2 (a) $3.5\mathbf{i} + 0.330\mathbf{j}$;   **(b)** 3.52 N;   **(c)** 3.52 N, 174.6° below **i**.

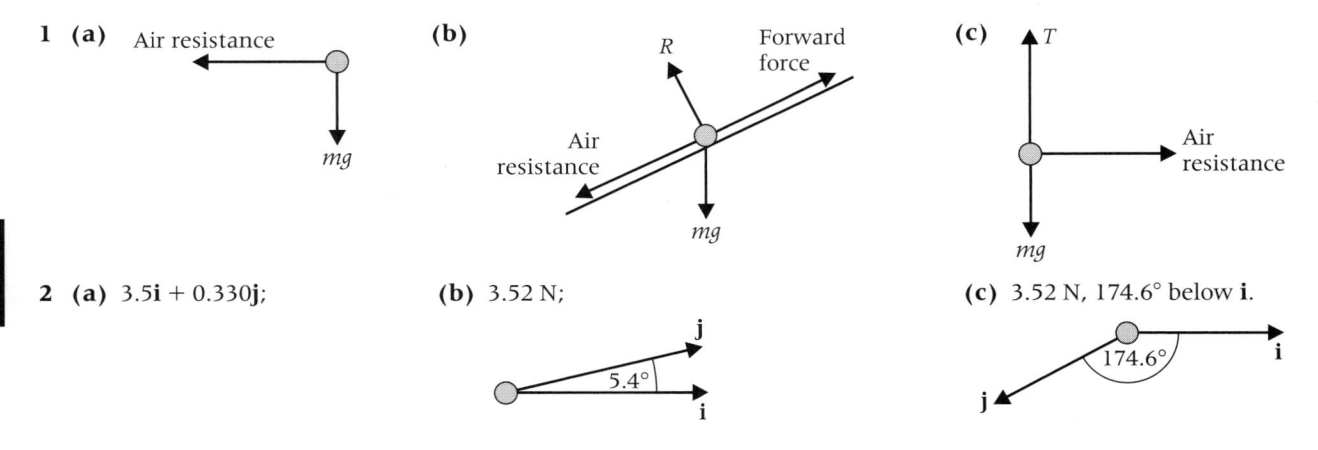

$5.4°$ $174.6°$

3 315 N, 375 N.

4 (a) 0.209;
(b) Nothing as 8 N is less than $F_{max} = 41$ N.

5 (a) 26.4 N;
(b) $\mu \geqslant 0.806$.

Learning objectives

After studying this chapter you should be able to:
- apply Newton's three laws of motion
- solve problems where the forces are in equilibrium
- solve problems where a particle is accelerating
- use Newton's third law.

5.1 Introduction

In this chapter we will examine particles in motion. The relationship between force and motion is described by Newton's laws. In the situations that we will encounter we will use Newton's laws to enable us to model motion produced by forces. If we were to study the motion of atomic particles it would be found that Newton's laws would not provide an accurate enough model and the theory of relativity should be considered.

5.2 Newton's first law

> A body will remain in a state of rest or will continue to move in a straight line with a constant velocity unless it is compelled to change that state by the action of a force.

In the previous chapter we considered the state of equilibrium to be when the resultant force is zero. Newton's first law implies that if a particle moves with a constant velocity then the resultant force will also be zero.

Worked example 5.1

A particle of mass 3 kg slides down a rough plane at an angle α to the horizontal at constant speed. If $\mu = 0.5$ find the angle α.

Solution

The diagram shows the forces acting on the particle.

As the particle is moving at a constant speed, the resultant force on the particle is zero.

Resolving horizontally: $3 \sin \alpha = F$

Resolving vertically: $3 \cos \alpha = R$

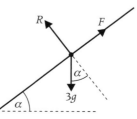

The laws of friction state that if sliding occurs on a rough surface then the frictional force has reached its maximum and $F = \mu R$, hence

$$F = 0.5R$$

$$3 \sin \alpha = 0.5 \times 3g \cos \alpha$$

$$\frac{\sin \alpha}{\cos \alpha} = 0.5$$

$$\tan \alpha = 0.5$$

$$\alpha = 26.6° \text{ (to three significant figures)}$$

Worked example 5.2

5

A stone of mass 100 grams is dropped in a lake. The stone experiences a resistive force which is proportional to the square of its speed, and reaches a maximum speed of 2 m s^{-1}. What will be the maximum speed of a similar stone of mass 50 grams?

Solution

As the stone falls it will increase in speed. The resistive force will therefore increase as well. The maximum speed will be achieved when the resistive force balances the weight of the stone, i.e. $R = mg$.

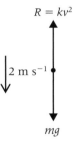

If R is proportional to v^2 then

$$R = kv^2, \text{ where } k \text{ is a constant.}$$

Hence at maximum speed

$$mg = kv^2$$

$$0.1 \times 9.8 = k \times 2^2$$

$$k = 0.245$$

The resistive force on the similar stone will be a function of its speed and shape, and not influenced by its mass. If we assume it has the same shape as the first stone then we can use $R = 0.245v^2$ again.

When this stone reaches its maximum speed

$$R = mg$$

$$0.05 \times 9.8 = 0.245v^2$$

$$v = 1.41 \text{ m s}^{-1} \text{ (to three significant figures)}$$

EXERCISE 5A

1 The following situations show a body moving with constant velocity in the direction shown, subject to unknown forces of magnitude X and Y. Find X and Y.

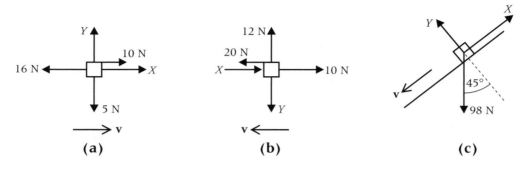

(a)   **(b)**   **(c)**

2 The following situations show a body moving with constant velocity in the direction shown, subject to an unknown force of magnitude F N acting an angle θ. Find F and θ.

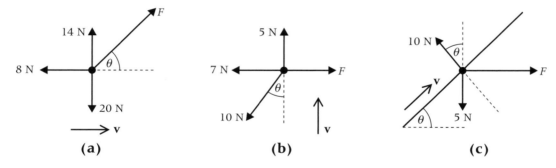

(a)   **(b)**   **(c)**

3 A lorry travels along a horizontal road with constant velocity. If the force which the engine exerts is 1350 N, what is the magnitude of the resistive force on the lorry?

4 A small object is being pulled across a horizontal surface at a constant velocity by a force of 12 N acting parallel to the surface. If the mass of the object is 5 kg, determine the coefficient of friction between the object and the surface.

5 A sledge of mass 16 kg is being pulled up the side of a hill of inclination 25°, at a constant velocity. The coefficient of friction between the sledge and the hill is 0.4, and the rope pulling the sledge exerts a force of magnitude T N at 15° to the hill.

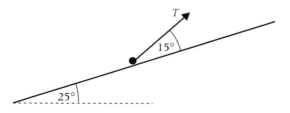

Find:

(a) the magnitude of the tension in the rope, and

(b) the magnitude of the normal reaction between the hill and the sledge.

6 A car of mass 1 tonne travels up a slope inclined at 3° to the horizontal, with constant velocity. If the engine exerts a force of 700 N, what is the magnitude of the resistive force on the car?

7 A lorry of mass 20 tonnes can travel up a slope inclined at 4° to the horizontal at a steady speed of 10 m s^{-1}, when the force produced by the engine is 15.2 kN. Assume that the air resistance is proportional to the square of the speed and ignoring other forms of resistance, find the maximum speed of the lorry when freewheeling down the same hill.

8 A lorry when fully laden has a mass of 40 tonnes. Its maximum speed when freewheeling down a slope inclined at 6° to the horizontal is 40 m s^{-1}, subject to air resistance which is proportional to the square of the speed of the lorry. When empty the lorry weighs 8 tonnes. What would be the maximum speed of the empty lorry when freewheeling down the same hill?

9 A skier of mass 80 kg can achieve a maximum speed of 35 m s^{-1} down an incline at an angle of 30° to the horizontal, subject to: air resistance which is proportional to her speed; and friction where the coefficient of friction between skier and the slope is 0.1. What will be the maximum speed of the skier down a slope inclined at an angle of 45° to the horizontal?

5.3 Newton's second law

Newton's second law describes the relationship between the change in motion of a body and the resultant force acting on the body. Provided the mass of the object concerned does not change, as it might if it was a space rocket, then the law can be stated mathematically as:

$$F = ma$$

where F is the resultant force on the body, which has mass m and a is its acceleration.

Note that:

(i) F and a are vector quantities. So the acceleration is in the direction of the resultant force.

(ii) The unit of mass is the kilogram. We measure acceleration in terms of m s^{-2}. A force of one newton (N) is defined as that force necessary to give a mass of 1 kg an acceleration of 1 m s^{-2}.

Worked example 5.3

Find the acceleration produced by a force of 20 N on a mass of 4 kg.

Solution

Using $F = ma$ with $F = 20$ and $m = 4$ gives:

$$20 = 4a$$
$$a = 5 \text{ m s}^{-2}$$

Worked example 5.4

Find the acceleration produced by the forces $(\mathbf{i} + 3\mathbf{j})$ N, $(3\mathbf{i} + \mathbf{j})$ N and $(4\mathbf{i} + 2\mathbf{j})$ N which act on a mass of 2 kg.

Solution

The resultant force is:

$$\mathbf{i} + 3\mathbf{j} + 3\mathbf{i} + \mathbf{j} + 4\mathbf{i} + 2\mathbf{j} = 8\mathbf{i} + 6\mathbf{j}.$$

Using $\quad \mathbf{F} = m\mathbf{a}$

$$2\mathbf{a} = 8\mathbf{i} + 6\mathbf{j}$$
$$\mathbf{a} = 4\mathbf{i} + 3\mathbf{j}$$

Note: the magnitude of the acceleration is $\sqrt{4^2 + 3^2} = 5$ m s^{-2}

Worked example 5.5

A particle of mass 3 kg is pulled across a rough horizontal plane, by a light inextensible string inclined at 30° to the horizontal. The tension in the string is 40 N. The coefficient of friction between the particle and the plane is 0.5. Find the acceleration of the particle.

Solution

The diagram shows the forces acting and the acceleration of the particle.

The particle accelerates along the plane so the resultant force must also be parallel to the plane. So the sum of the vertical components of the forces must be zero.

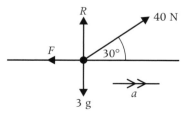

Resolving vertically:

$$R + 40 \sin 30° - 3g = 0$$
$$R = 9.4 \, \text{N}$$

The frictional force will be at its maximum, i.e.

$$F = \mu R = 0.5 \times 9.4$$
$$F = 4.7 \, \text{N}$$

Using $F = ma$ parallel to the plane:

$$40 \cos 30° - 4.7 = 3a$$
$$a = 9.98 \, \text{m s}^{-2} \quad \text{(to three significant figures)}$$

Worked example 5.6

A stone is projected with speed $10 \, \text{m s}^{-1}$ down the line of greatest slope of a plane inclined at $30°$ to the horizontal. The coefficient of friction between the slope and the stone is 0.8. Find the distance travelled by the stone before coming to rest.

Solution

The diagram shows the forces acting and the acceleration of the particle. The first step is to find the acceleration of the stone.

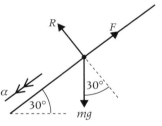

The components of the forces perpendicular to the plane must be in equilibrium. So resolving perpendicular to the plane gives

$$R = mg \cos 30°.$$

Friction is limiting, i.e.

$$F = \mu R$$
$$= 0.8 \, mg \cos 30°$$

Using Newton's second law parallel to the plane gives:

$$mg \sin 30° - F = ma$$
$$mg \sin 30° - 0.8 \, mg \cos 30° = ma$$

The mass cancels throughout this equation and we get:

$$a = g \sin 30° - 0.8g \cos 30°$$
$$= -1.89 \, \text{m s}^{-2} \quad \text{(to three significant figures)}$$

Throughout this motion all the forces are constant so the acceleration will also be constant. If the acceleration is constant we can use the constant acceleration formulae.

Using

$$v^2 = u^2 + 2as$$

gives

$$0^2 = 10^2 + 2 \times (-1.89)d$$

where d is the distance the stone slides. This gives

$$d = 26.5 \, \text{m}$$

EXERCISE 5B

1 A single force of magnitude 8 N acts on a particle of mass 16 kg. Find the acceleration produced.

2 The acceleration of a particle of mass 2 kg is 3 m s^{-2}. Find the magnitude of the resultant force on the particle.

3 Forces of $(\mathbf{i} + \mathbf{j})$ N, $(3\mathbf{i} - 2\mathbf{j})$ N and $(\mathbf{i} + 3\mathbf{j})$ N act on a particle of mass 2 kg. Find the acceleration of the particle in the form $a\mathbf{i} + b\mathbf{j}$.

4 A crane lifts a load of mass 2 tonnes vertically from rest. The tension in the cable is 21 kN. Find the acceleration of the load.

5 A package with mass 300 kg is lifted vertically upwards. Find the tension in the cable which lifts the package, when the package:

   (a) accelerates upwards at 0.1 m s^{-2}

   (b) accelerates downwards at 0.2 m s^{-2}

   (c) travels upwards with a retardation of 0.1 m s^{-2}.

6 Find the unknown accelerations, forces and angles in the following situations.

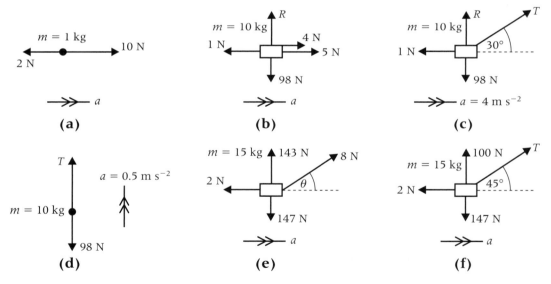

7 A child slides down a steep, straight slide that is inclined at 60° to the horizontal. The child has mass 30 kg and the coefficient of friction between the slide and the child is 0.6. Assume that there is no air resistance.

   (a) Draw a diagram to show the forces acting on the child, while sliding down the slide.

   (b) Calculate the magnitude of the normal reaction force on the child.

(c) Show that the magnitude of the friction force that acts on the child is 88.2 N.

(d) Calculate the acceleration of the child.

(e) What modelling assumption have you made about the child in your solution?   [A]

8 Kate and Joe are playing on a children's slide. The slide is straight and inclined at an angle of 30° to the horizontal, as shown in the diagram. Kate's mass is 25 kg and the coefficient of friction between Kate and the slide is 0.25.

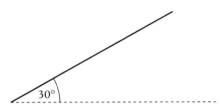

(a) Kate slides down the slide.
  (i) Show that the frictional force between her and the slide is approximately 53 N.
  (ii) Find her acceleration.

(b) On one occasion, before Kate slides down, she is held at rest on the slide by Joe. He exerts a force on her which is parallel to the slide.
  (i) Find the minimum possible value of this force.
  (ii) Joe then releases Kate and she slides down. Find her speed after sliding 5 metres from her release point.   [A]

9 A car of mass 1 tonne travels along a horizontal road and brakes from 50 m s^{-1} to rest in a distance of 300 m. Find the braking force on the car.

10 A particle of mass m kg slides down a smooth plane, inclined at an angle of 30° to the horizontal. Find the acceleration of the particle down the plane.

11 A particle of mass 6 kg starts from rest and accelerates uniformly. The resultant force on the particle has magnitude 15 N. Find the time taken to reach a speed of 10 m s^{-1}.

12 A particle of mass 20 kg is pulled across a rough horizontal plane by a light inextensible string, inclined at 30° to the horizontal. If the tension in the string is 50 N and the acceleration produced is 0.5 m s^{-2} find the frictional force on the particle and the coefficient of friction.

13 A particle of mass m kg slides down a smooth inclined plane with an acceleration of 2 m s^{-2}. Find the inclination of the plane to the horizontal.

14 A cyclist travels up a slope inclined at 4° to the horizontal. The cyclist and cycle are modelled as a particle of mass 80 kg. A constant air resistance force of magnitude 20 N acts throughout the motion. The diagram shows the forces that are assumed to act on the particle.

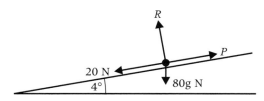

(a) Find the value of P:
    (i) when the cyclist moves at a constant speed,
    (ii) when the cyclist accelerates at 0.2 m s^{-2}.

The cyclist accelerates at 0.2 m s^{-2} from rest until her speed reaches 4 m s^{-1}. She then travels at a constant speed.

(b) **(i)** Find the time for which the cyclist accelerates.
    (ii) Sketch a velocity–time graph for the motion of the cyclist.

(c) **(i)** Explain why modelling the air resistance as a constant might not be appropriate.
    (ii) Sketch a velocity–time graph to show how a more realistic air resistance would affect the motion of the cyclist. [A]

15 A car moves along a straight road. When it passes a set of traffic lights, the car is travelling at a speed of 8 m s^{-1}. The car then moves with constant acceleration for 10 seconds and travels 200 metres.

(a) Show that the acceleration of the car is 2.4 m s^{-2}.

(b) Find the speed of the car at the end of the 10 seconds.

(c) The road is horizontal and the car has mass 1200 kg. A constant resistance force of 1800 N acts on the car while it is moving.
    (i) Find the magnitude of the driving force that acts on the car while it is accelerating.
    (ii) At the end of the 10 second period the driving force is removed. The car then moves subject to the resistance force of 1800 N until it stops. Find the distance that the car travels while it is slowing down. [A]

16 A particle of mass 500 grams starts from rest at the point A which has position $(4\mathbf{i} + 2\mathbf{j})$ m. The resultant force on the particle is $(\mathbf{i} + 2\mathbf{j})$ N. Find the position vector of the particle after 3 seconds.

17 Two tugs are towing a large oil tanker into harbour. Tug A's engines can produce a pulling force of 80 kN while tug B's engines can produce 65 kN of force.

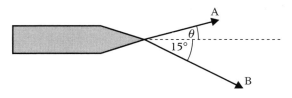

(a) Calculate the angle θ necessary for the tanker to move directly forwards.

(b) Given that there is a resistance to the motion of the tanker of 25 kN directly opposing motion, find the magnitude of the resultant force on the tanker to the nearest 1000 N. Find also the acceleration of the tanker if it has a mass of 20 000 tonnes.

18 A sledge of mass 30 kg is accelerating down a hill while a boy is trying to prevent it from sliding by pulling on a rope attached to the sledge with a force of 40 N. The hill has inclination 28° and the rope is inclined to the hill at 10°. The coefficient of friction between the sledge and the hill is 0.35. Find:

(a) the magnitude of the normal reaction force on the sledge,

(b) the resultant force on the sledge acting down the hill,

(c) the magnitude of the sledge's acceleration.

19 A car, of mass 900 kg, is initially at rest. On a short journey the car
I. accelerates uniformly for T seconds to a speed of 20 m s^{-1},
II. then travels at this speed for a period of time,
III. then decelerates uniformly for 2T seconds before coming to rest.

(a) In one journey the car moves for a total of 40 seconds and travels a total of 620 m. Using this information:
  (i) Sketch a velocity–time graph and hence, or otherwise, find T.
  (ii) Calculate the magnitude of the resultant force on the car during each stage of the journey.
  (iii) Sketch a graph to show how the resultant force acting on the car varies with time.
  (iv) Find the speed of the car after it has travelled 20 m.

(b) In the case when $T = 5$, find the time that it would take the car to complete a 1000 m journey.   [A]

20 A tree trunk, of mass 250 kg, is pulled up a slope by a chain attached to a tractor. The chain is at an angle of 10° to the slope. The slope itself is at 8° to the horizontal. The tree trunk initially accelerates at 0.2 m s^{-2}. A friction force, of magnitude 2000 N, acts on the tree trunk.

   (a) Model the tree trunk as a particle. Draw and label a diagram to show the forces acting on it.

   (b) Find the initial tension in the chain.

   (c) Explain why the tension in the chain will probably decrease.   [A]

21 A ball is thrown vertically upwards from ground level. Throughout its motion it is acted on by gravity and a resistance force of constant magnitude.

   The ball reaches a maximum height of 1.5 metres after 0.5 seconds.

   (a) Find:
      (i) the initial speed of the ball,
      (ii) its acceleration as it is moving upwards.

   The mass of the ball is 0.2 kg.

   (b) Show that the magnitude of the resistance force is 0.44 N.

   (c) Find the acceleration of the ball as it falls back to the ground.

   (d) Find the total time that the ball is in the air.   [A]

22 Two cables, *AB* and *AC*, are attached to a cable car, as shown in the diagram. The cable car has mass 450 kg.

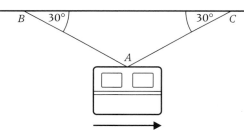

   The cable car travels horizontally in the direction shown by the arrow. Model the cable car as a particle and assume that there is no air resistance present. As the cable car moves, the angles shown in the diagram do not change.

   (a) The cable car travels at a constant speed. Show that the tension in each cable is 4410 N.

   (b) The cable car now accelerates in the direction of the arrow at 0.5 m s^{-2}. Find the tension in each cable.

   (c) Describe how your answers to **(b)** would change if air resistance was taken into account.   [A]

23 A child is sliding at a constant speed of $4\,\text{m}\,\text{s}^{-1}$ down a long slide. The child has a mass of 45 kg. The slide is inclined at an angle of 40° to the horizontal. Assume that a constant friction force, of magnitude 89 N, acts on the child.

(a) Use the data given to explain why air resistance must be taken into account when modelling the motion of the child. Find the magnitude of the air resistance acting on the child, when he is travelling at a constant speed of $4\,\text{m}\,\text{s}^{-1}$.

(b) Assume that the magnitude of the air resistance is proportional to the speed of the child. The next time that he uses the slide he starts from rest and accelerates. Find his acceleration, when he is moving at $1\,\text{m}\,\text{s}^{-1}$. [A]

24 A sledge, of mass 12 kg, is pulled up a rough slope which is inclined at an angle of 10° to the horizontal. The coefficient of friction between the slope and the sledge is 0.2.

(a) The sledge is pulled by a rope that is parallel to the slope, as shown in the diagram.

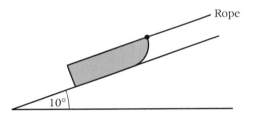

(i) Draw a diagram to show the forces acting on the sledge.

(ii) Find the magnitude of the normal reaction force acting on the sledge.

(iii) Given that the acceleration of the sledge is $0.5\,\text{m}\,\text{s}^{-2}$, show that the tension in the rope is approximately 50 N.

(b) The sledge is then pulled with the rope at an angle of 30° to the slope, as shown in the diagram.

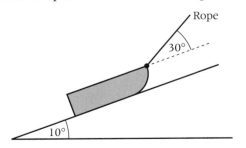

Find the acceleration of the sledge if the tension in the rope is 60 N.

(c) Write down **two** modelling assumptions that you have made. [A]

5.4 Newton's third law

Newton's third law states:

> For every action there is an equal but opposite reaction.

This law is often misunderstood, but is really very simple. The best way to understand it is to consider some simple examples.

If you stand on a table, your feet push down on the table. In response to this the table exerts an upward force on your feet. This upward force is the equal but opposite reaction to the downward force you exert on the table. These are both normal reaction forces.

The diagrams show these normal reaction forces (R).

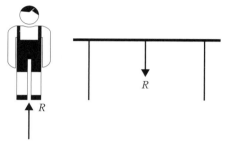

It is important to note that this example does not directly involve the weight of the person (W), although $W = R$.

A second example concerns the orbit of the Moon around the Earth. The Earth exerts a gravitational force on the Moon and the Moon exerts an equal but opposite force on the Earth. The effect of the force acting on the Moon is to keep it in orbit around the Earth. The effect of the force on the Earth is to cause the tides. The important idea, however, is that both exert forces of equal size on each other, as shown in the diagram.

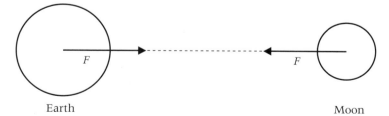

Earth   Moon

As a final example consider a car towing a caravan. The car exerts a forward force on the caravan, but the reaction to this is the backwards force that the caravan exerts on the car.

EXERCISE 5C

1 Use Newton's third law to explain why you would hurt your hand if you punched a hard object with it.

2 A ladder leans against a smooth wall.

(a) Draw diagrams to show the force that the top of the ladder exerts on the wall and the force that the wall exerts on the top of the ladder.

(b) Use Newton's third law to explain why these forces have equal magnitudes.

3 A child jumps off a table and lands on the ground. Describe how the force that the ground exerts on the child varies. Also describe how the force that the child exerts on the ground varies.

4 A man, of mass 78 kg, stands in a lift of mass 200 kg that is accelerating upwards at 0.5 m s^{-2}. Calculate the magnitudes of the forces that act on the lift. Also draw a diagram to show how they act.

5 Three carriages are coupled to an engine on a set of railway lines. The carriages and engine move forward on horizontal tracks. Draw diagrams to show the forces acting on each of the carriages. Clearly show any forces that have the same magnitude.

Key point summary

Formula to learn:

$$F = ma$$

1 If a body is at rest or moves with a constant velocity the forces acting on it must be in equilibrium. *p84*

2 When applying Newton's second law, remember that F represents the resultant force. *p87*

3 When two bodies interact, the force exerted by the first body on the second body is equal and opposite to the force exerted by the second body on the first. *p96*

Test yourself	What to review
1 A helicopter of mass 880 kg is rising vertically at a constant rate. Find the magnitude of the lift force acting on the helicopter. How would your answer change if the helicopter was descending at a constant rate?	*Section 5.2*
2 A child, of mass 30 kg, slides down a slide at a constant speed. Assume that there is no air resistance acting on the child. The slide makes an angle of 40° with the horizontal. Find the magnitude of the friction force on the child and the coefficient of friction.	*Section 5.2*
3 A lift and its passengers have a total mass of 300 kg. Find the tension in the lift cable if: **(a)** it accelerates upwards at 0.2 m s^{-2} **(b)** it accelerates downwards at 0.05 m s^{-2}.	*Section 5.3*
4 A van, of mass 1200 kg, rolls down a slope, inclined at 3° to the horizontal and experiences a resistance force of magnitude 400 N. Find the acceleration of the van.	*Section 5.3*

Test yourself ANSWERS

4 0.180 m s^{-2}.

3 (a) 3000 N; **(b)** 2925 N.

2 189 N, 0.839.

1 8624 N, no change.

CHAPTER 6

Connected particles

Learning objective

After studying this chapter you should be able to:
- solve problems of motion involving connected particles.

6.1 Introduction

In this chapter we will study the motion of bodies that are connected in some way. This will focus mainly on objects that are connected by a light, inextensible string, but we will also look at some other similar examples, for example when a car tows a caravan.

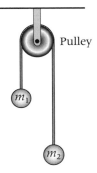

Pulley

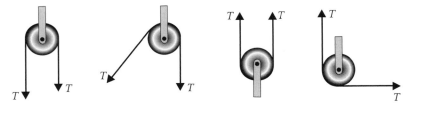

6.2 Motion of two connected particles

When two particles are connected by a light string, it is important to realise that the tension will have the same magnitude throughout the string, although the tension may act in different directions. This understanding of the tension in a string is the key to solving problems that involve connected particles.

We will only consider cases where the string joining the two objects remains taut, so that both objects have the same acceleration.

When attempting problems with connected particles it is important to consider each particle separately and draw a force diagram for each particle.

Worked example 6.1

Two particles, of mass 3 kg and 7 kg, are connected by a light, inextensible string. The string passes over smooth pulley, as shown in the diagram.

Find the acceleration of the masses and the tension in the string.

Solution

First consider the 3 kg mass. Two forces act on this mass, its weight downwards and the tension upwards.

The resultant of these two forces is $T - 3g$.

Applying Newton's second law ($F = ma$) gives the equation:

$$T - 3g = 3a \qquad [1]$$

where a is the acceleration of the masses.

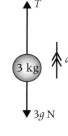

Now consider the 7 kg mass. Similarly the resultant force on this particle is $7g - T$ and applying Newton's second law gives the equation:

$$7g - T = 7a \qquad [2]$$

These two equations are a pair of simultaneous equations and can now be solved.

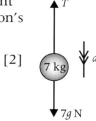

From equation [1] $T = 3a + 3g$. This can then be used to substitute for T in equation [2] to give:

$$7g - (3a + 3g) = 7a$$

$$4g = 10a$$

$$a = \frac{4g}{10}$$

$$= 3.92 \text{ m s}^{-2}$$

This acceleration can then be substituted into $T = 3a + 3g$ to give T.

$$T = 3a + 3g$$

$$= 3 \times 3.92 + 3 \times 9.8$$

$$= 41.16 \text{ N}$$

Worked example 6.2

The diagram shows a particle, of mass
7 kg, sliding on a rough horizontal
surface and connected by a light,
inextensible string to another particle,
of mass 5 kg. The coefficient of friction
between the surface and the particle is
0.2. Find the acceleration of each particle
and the tension in the rope.

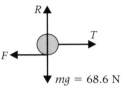

Solution

First consider the particle on the surface. The weight of the
particle is 68.6 N, which is balanced by the normal reaction R, so
that $R = 68.6$ N.

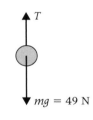

As the particle is sliding the friction force takes its maximum
value of μR, so that $F = 0.2 \times 68.6 = 13.72$ N.

So the resultant force, to the right, on this particle is:

$$T - F = T - 13.72$$

Using Newton's second law, $F = ma$, gives:

$$T - 13.72 = 7a \qquad [1]$$

Now consider the suspended particle. The resultant downward
force on this particle is:

$$49 - T$$

Using Newton's second law, $F = ma$, gives:

$$49 - T = 5a \qquad [2]$$

We now have a pair of simultaneous equations that need to be
solved.

From equation [1], $T = 7a + 13.72$, and substituting this into
equation [2] gives:

$$5a = 49 - (7a + 13.72)$$

$$12a = 49 - 13.72$$

$$a = \frac{35.28}{12} = 2.94 \text{ m s}^{-2}$$

This can now be used to find T, by substituting for a in the
equation:

$$T = 7a + 13.72.$$

$$T = 7 \times 2.94 + 13.72$$

$$= 34.3 \text{ N}$$

Worked example 6.3

A truck is pulled up a slope by a rope that passes over a pulley and is attached to a large barrel of water, which hangs over the edge of a cliff. The slope is at an angle of 30° to the horizontal and the mass of the truck and its contents is 500 kg. Assume that a constant resistance force of 100 N acts on the truck. The barrel and water have a total mass of 1000 kg.

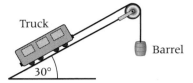

(a) State any assumptions that you need to make.

(b) Find the acceleration of the truck and the tension in the rope.

Solution

(a) The following assumptions should be made:
 - the rope is light and inextensible;
 - there is no air resistance;
 - there is no friction in the pulley;
 - the barrel and truck can be modelled as particles;
 - the truck moves up the slope.

(b) Consider first the forces acting on the barrel.

The resultant force is $1000g - T$, so applying Newton's second law gives the equation below.

$$1000g - T = 1000a \qquad [1]$$

Now consider the forces on the truck. Resolving parallel to the slope gives the resultant force as:

$$T - 100 - 500g \sin 30° = T - 100 - 250g$$

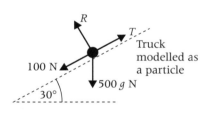

Applying Newton's second law gives:

$$T - 100 - 250g = 500a \qquad [2]$$

We now have a pair of simultaneous equations that can be solved.

From equation [1] $T = 1000g - 1000a$.

This can be substituted into equation [2] to give:

$$T - 100 - 250g = 500a$$

$$1000g - 1000a - 100 - 250g = 500a$$

$$750g - 100 = 1500a$$

$$a = \frac{750g - 100}{1500}$$

$$= \frac{29}{6}$$

$$= 4.83 \text{ m s}^{-2}$$

Now the value for a can be substituted into $T = 1000g - 1000a$, to find the tension.

$$T = 1000g - 1000a$$
$$= 1000 \times 9.8 - 1000 \times \frac{29}{6}$$
$$= 4967 \text{ N}$$

Worked example 6.4

A car of mass 1100 kg pulls a caravan of mass 900 kg. The action of the engine causes a force of magnitude 3000 N to act on the car. Resistance forces of magnitude 300 N and 450 N act on the car and caravan, respectively.

Find the acceleration of the car and caravan and the force that the car exerts on the caravan.

Solution

First consider the car and caravan as a single particle. The resultant force is $3000 - 300 - 450 = 2250$.

Applying Newton's second law gives:

$$2250 = 2000a$$
$$a = 1.125 \text{ m s}^{-2}$$

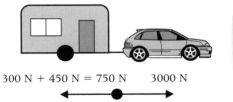

300 N + 450 N = 750 N   3000 N

Car and caravan modelled as a particle

Now consider the caravan.

The resultant force is $T - 450$, so applying Newton's second law gives the equation:

$$T - 450 = 900 \times 1.125$$
$$T = 1463 \text{ N}$$

450 N   T

Forces on the caravan

EXERCISE 6A

1 Two particles, of mass 3 kg and 5 kg, respectively, are attached to the ends of a light, inextensible string that passes over a smooth pulley. Both strings are vertical. Find the acceleration of the particles and the tension in the string.

2 The diagrams below show particles that are connected by a light, inextensible string that passes over a smooth pulley. In each case find the acceleration of the particles and the tension in the string.

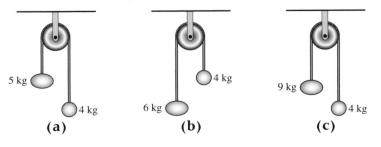

5 kg   4 kg   4 kg   6 kg   9 kg   4 kg

(a)   (b)   (c)

3 Two particles, of mass 4 kg and 7 kg, are attached to the ends of a light, inextensible string, which passes over a smooth pulley. Initially the particles are at rest and at the same level. The particles are then released.

   (a) Find the acceleration of the particles.

   (b) Find the time when the difference in height of the particles is 1 m.

4 Two particles of mass m and $\dfrac{3m}{2}$ are connected by a light, inextensible string that passes over a smooth pulley. Find the acceleration of the particles and the tension in the string in terms of m and g.

5 The diagram shows a block of mass 2 kg, that slides on a rough horizontal table. It is attached by a light, inextensible string to a 3 kg mass. If the block accelerates at 4.9 m s^{-2}, find the coefficient of friction between the block and the table.

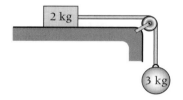

6 A block, of mass 6 kg, rests on a rough, horizontal surface. The coefficient of friction between the block and the surface is 0.2. A light, inextensible string attached to the block passes over a smooth pulley. A weight, of mass 2 kg, hangs from the other end of the string, as shown in the diagram below.

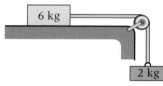

Find the tension in the string and the acceleration of the block.   [A]

7 The diagram shows a mass A of 5 kg initially at rest on a horizontal table. A resistance force of 10 N acts against the motion of A which is connected to mass B of 3 kg by a light, inextensible string which passes over a smooth pulley. The system is released from rest.

   (a) Calculate the acceleration of A.

   (b) Calculate the tension in the string.

   After a short time, B reaches the floor.

   (c) Calculate the acceleration of A now.

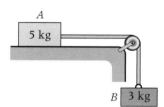

8 A breakdown truck tows a car of mass 1200 kg. Assume that the rope is horizontal and that the truck moves on a horizontal surface. Calculate the tension in the tow rope if the car is:

   (a) accelerating at 0.5 m s^{-2} and experiencing a resistance force of 500 N;

   (b) travelling at constant speed but experiencing a resistance force of 400 N.

9 Two bodies A and B of mass 4 kg and 2 kg, respectively, are attached by a light, inextensible string passing over a smooth pulley. A rests on a table and B hangs over the side. Resistance forces on A amount to 8 N.

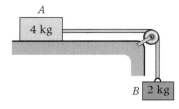

The system is released from rest. Calculate the acceleration of the system and the tension in the string. Find the speed of B when it has fallen 2 m.

10 Two particles, of masses 4 kg and 6 kg, are connected by a light, inextensible string that passes over a smooth, light pulley. The two particles are released from rest, with the string taut, as shown in the diagram.

(a) Show that the acceleration of each particle is 1.96 m s^{-2}.

(b) Calculate the tension in the string. [A]

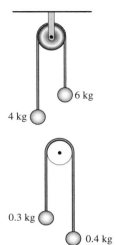

11 Two particles, of masses 0.3 kg and 0.4 kg, are connected by a light, inextensible string which hangs over a smooth fixed peg, as shown in the diagram. The system is released from rest.

(a) **(i)** Show that, in the subsequent motion, the acceleration of the particles is of magnitude 1.4 m s^{-2}.

(ii) Find the tension in the string during this motion.

(b) Find the distance travelled by each particle during the first 2 seconds of the motion. [A]

12 A box, of mass 6 kg, is at rest on a rough horizontal table. A light, inextensible string is attached to the box. This string passes over a smooth, light pulley and the other end is attached to a sphere of mass 4 kg, as shown in the diagram. Assume that the string is horizontal between the box and the pulley. The coefficient of friction between the box and the table is 0.5.

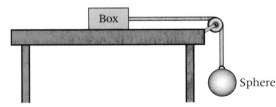

The sphere is released from rest, with the string taut, and the box slides along the table.

(a) Calculate the magnitude of the friction force acting on the box.

(b) Show that the acceleration of the system is 0.98 m s^{-2}.

(c) Calculate the tension in the string. [A]

13 Two particles are connected by a light, inextensible string, which passes over a smooth, light pulley, as shown in the diagram. The particle A, of mass 0.5 kg, is in contact with a rough horizontal surface, and the particle B, of mass 0.2 kg, hangs freely. The coefficient of friction between A and the surface is $\frac{2}{7}$.

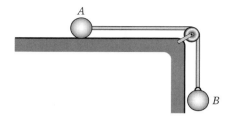

The system is released from rest with the string taut and A moves towards the pulley.

(a) Show that the frictional force between A and the surface is of magnitude 1.4 N.

(b) Find the acceleration of the particles.

(c) Find the tension in the string.

(d) Find the time taken for the particles to travel 0.625 metres, given that A has not then reached the pulley. [A]

14 A block, of mass 6 kg, is held at rest on a rough horizontal table. The block is attached, by a light string that passes over a light, smooth pulley, to a sphere of mass 4 kg, that hangs freely, as shown in the diagram.

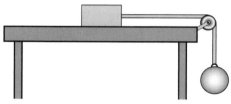

(a) The block is released and travels 60 cm in 2 seconds. Show that the acceleration of the block is 0.3 m s^{-2}.

(b) Find the magnitude of the tension in the string.

(c) Find the magnitude of the friction force that acts on the block.

(d) Find the coefficient of friction between the block and the table. Give your answer correct to two significant figures. [A]

15 Two particles, P and Q, are connected by a light, inextensible string which passes over a smooth, fixed peg, as shown in the diagram.

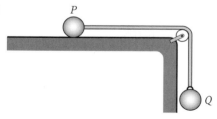

The particle P is of mass $2m$ and Q is of mass m. The particle P is in contact with a horizontal surface, which may be modelled as either rough or smooth.

(a) In the first case, the surface is modelled as rough and the particle Q hangs at rest. The coefficient of friction between P and the surface is μ.

    (i) Find the tension in the string.

    (ii) Find the range of possible values of μ.

(b) In the second case, the surface is modelled as smooth. The system is released from rest with the string taut and the particle P moves towards the peg.

    Find the tension in the string. [A]

16 A builder uses a rope that passes over a pulley to raise and lower concrete blocks while building a house. The diagram shows the initial positions of two loads of the blocks.

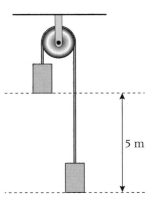

Each load consists of eight blocks, each of mass 5 kg. Initially the lower load is resting on the ground. When two blocks are accidentally removed from the lower load the system begins to move.

5 m

(a) Making any necessary modelling assumptions:
 (i) show that the acceleration of the loads is $1.4\,\text{m s}^{-2}$,
 (ii) find the tension in the rope.

(b) State two assumptions made in **(a)**.

(c) Find the speed of the two loads when they are at the same level. [A]

17 Two children are holding the ends of a light, inextensible rope, which passes over a light, smooth pulley. Initially Tom, who has a mass of 40 kg, is standing at ground level and Simon, who has a mass of 60 kg, is on the edge of a fixed platform 2 metres above ground level. Model the two boys as particles, one initially at ground level, and the other initially at a height of 2 metres. The rope is taut.

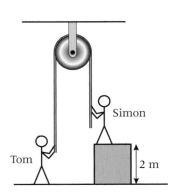

Simon

Tom

2 m

Simon steps off the platform and as he falls vertically, Tom rises vertically.

(a) Assume that the rope remains taut while the boys are moving.
 (i) Show that the acceleration of each boy is $1.96\,\text{m s}^{-2}$.
 (ii) Find the tension in the rope.

(b) Find the total distance that Tom travels upwards. [A]

18 The diagram shows a car pulling a trailer in a straight line on a horizontal stretch of road.

The mass of the car is 1250 kg and the total resistance force acting on the car is 500 N.

The mass of the trailer is 250 kg and the total resistance force acting on the trailer is 100 N.

(a) During part of the journey, a constant braking force is applied to the car, causing the car and trailer to decelerate at a constant rate of $0.5\,\text{m s}^{-2}$.
 (i) By considering the forces on the trailer, find the magnitude of the force in the towbar between the car and the trailer.
 (ii) Find the magnitude of the braking force applied to the car.

(b) Later in the journey, the car and trailer travel with constant speed.
 State the magnitude of the tension in the towbar. [A]

19 A particle, of mass M, is on a rough horizontal table. A light, inextensible string is attached to this particle and passes over a smooth pulley. Attached to the other end of the string, so that it hangs vertically, is another particle of mass m. The coefficient of friction between the particle and the table is μ. The particles are initially at rest.

   (a) Show that the acceleration of the particles is $\dfrac{g(m - \mu M)}{m + M}$.

   (b) Find the tension in the string.

   (c) Find the time that it takes the particle on the table to move 0.5 m and its speed at this time.

20 The diagram shows two particles connected by a light, inelastic rope. One has a mass of 4 kg and is on a smooth slope and the other has a mass of 6 kg and hangs freely.

   (a) Find the acceleration of the particles if $\alpha = 30°$.

   (b) Find the tension in the rope if $\alpha = 10°$.

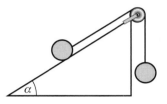

21 A car of mass 1200 kg is to be pulled 30 m up a slope inclined at 5° to the horizontal. A rope is attached to the car and passes over a smooth pulley and hangs vertically in an old mine shaft. A concrete weight of mass 200 kg is attached to the other end of the rope and released from rest. Assume that there is no resistance to the motion of the car.

   (a) Find the acceleration of the car.

   (b) How long does the car take to travel the 30 m?

22 At an old mine trolleys move on tracks up a slope inclined at 10° to the horizontal. They are pulled up by a rope that passes over a pulley and then hangs down the mine shaft. A weight of mass 250 kg is attached to this end of the rope. The mass of an empty truck is 800 kg. The truck is subject to a resistance force of 100 N as it moves.

   (a) Find the acceleration of an empty truck as it moves up the slope.

   (b) At the top of the slope the truck is loaded with 1000 kg of coal. Find the acceleration of the truck as it moves down the slope.

23 Two particles are connected by a light string that passes over a smooth, light pulley as shown in the diagram.

   The 4 kg particle is on a smooth, fixed slope, which is at an angle of 60° to the horizontal. The 3 kg particle hangs with the string vertical.

   The particles are released from rest at the position shown.

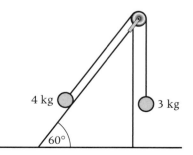

   (a) Show that the acceleration of the particles is approximately 0.65 m s^{-2}.

   (b) By considering the 3 kg particle, determine the tension in the string.   **[A]**

Key point summary

1 Remember that the tension is constant throughout the string. *p99*

2 The magnitude of the acceleration of both particles is the same. *p99*

3 Consider each particle separately, applying Newton's second law to each in turn. *p99*

Test yourself	What to review

1 Particles of mass 4 kg and 3.5 kg are attached to the ends of a light, inextensible string, which passes over a smooth pulley. The system is released from rest. *Section 6.2*

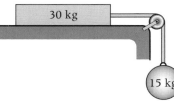

3.5 kg   4 kg

(a) Find the tension in the string.

(b) Find the time that it takes for the heavier particle to fall 0.5 m.

2 A block, of mass 30 kg, rests on a rough horizontal surface. The coefficient of friction between the surface and the block is 0.4. A light inextensible rope is attached to the block and passes over a smooth pulley. A particle, of mass 15 kg, hangs from the other end of the rope. The system is released from rest. *Section 6.2*

30 kg

15 kg

(a) Find the acceleration of the block and the tension in the rope.

(b) Find the speed of the block when it has travelled 40 cm.

Test yourself ANSWERS

1 **(a)** 36.6 N;   **(b)** 1.24 s.

2 **(a)** 0.653 m s^{-2}, 137.2 N;   **(b)** 0.723 m s^{-1}.

CHAPTER 7

Projectiles

Learning objectives

After studying this chapter you should be able to:
- find the time of flight of a projectile
- find the range of a projectile
- find the maximum height of a projectile
- modify equations to take account of height of release.

7.1 Introduction

In this chapter we will study the motion of objects that move only under the influence of gravity. For example we could consider the paths of a golf ball, a tennis ball, a football or a stunt motorcyclist, while they are in the air.

7.2 A model for the motion of a projectile

We make a number of assumptions that allow us to model these situations. The key assumptions are listed below:

- The object that is moving is modelled as a particle.
- Gravity is the only force that acts, so that we ignore forces such as air resistance and lift that may be present in reality.
- The object is set in motion so that it has a specific initial velocity.

To develop a mathematical model for a projectile we will first consider the motion of a particle that is launched on horizontal ground and is initially at ground level.

> As the particle moves only under the influence of gravity it will have a constant acceleration of g m s^{-2}. If we define unit vectors $\mathbf{i}$ and $\mathbf{j}$ as horizontal and vertical, respectively, we can write the acceleration as
>
> $$\mathbf{a} = -g\mathbf{j}.$$

We need to be able to describe the initial velocity of the particle. If it initially moves at V m s^{-1}, at an angle θ above the horizontal, then the initial velocity of the particle will be

$$\mathbf{u} = V\cos\theta\,\mathbf{i} + V\sin\theta\,\mathbf{j}$$

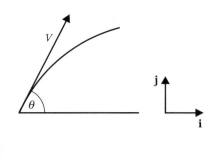

Using the constant acceleration equations

$$\mathbf{v} = \mathbf{u} + \mathbf{a}t \quad \text{and} \quad \mathbf{r} = \mathbf{u}t + \frac{1}{2}\mathbf{a}t^2$$

allows us to formulate expressions for the velocity and position of the projectile at time t.

$\mathbf{v} = \mathbf{u} + \mathbf{a}t$

$\phantom{\mathbf{v}} = V\cos\theta\,\mathbf{i} + V\sin\theta\,\mathbf{j} - gt\mathbf{j}$

$\phantom{\mathbf{v}} = V\cos\theta\,\mathbf{i} + (V\sin\theta - gt)\mathbf{j}$

and

$\mathbf{r} = \mathbf{u}t + \dfrac{1}{2}\mathbf{a}t^2$

$\phantom{\mathbf{r}} = (V\cos\theta\,\mathbf{j} + V\sin\theta\,\mathbf{j})t - \dfrac{1}{2}gt^2\mathbf{j}$

$\phantom{\mathbf{r}} = V\cos\theta\,t\,\mathbf{i} + \left(V\sin\theta\,t - \dfrac{1}{2}gt^2\right)\mathbf{j}$

The diagram below shows the typical path of a projectile.

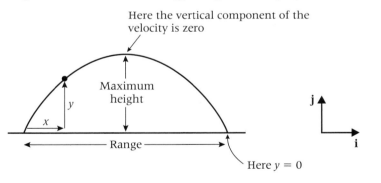

Here the vertical component of the velocity is zero

Maximum height

y

x

Range

Here $y = 0$

When working with projectile problems it is often necessary to consider either the horizontal or vertical components of the position. These can be written as

$$x = V\cos\theta\,t \quad \text{and} \quad y = V\sin\theta\,t - \frac{1}{2}gt^2$$

A great many projectile problems can be solved by starting with these two equations. You should learn these formulae so that you can apply them quickly and easily. Their use is illustrated in the following examples.

Worked example 7.1

A football is kicked from ground level so that its initial speed is 20 m s^{-1} and at an angle of $30°$ above the horizontal. Assume that the ball moves only under the influence of gravity.

(a) Find the time of flight of the ball.

(b) Find the range of the ball.

(c) The maximum height of the ball.

Solution

The position of the ball, at time t seconds, is given by:

$$x = V \cos \theta t \qquad\qquad y = V \sin \theta t - \frac{1}{2}gt^2$$

$$= 20 \cos 30°t \quad \text{and} \quad = 20 \sin 30°t - \frac{1}{2} \times 9.8t^2$$

$$= 10t - 4.9t^2$$

(a) The time of flight is the time that the ball is in the air, from the time it is kicked to the time when it first hits the ground. This can be found by finding the times when $y = 0$.

$$10t - 4.9t^2 = 0$$

$$t(10 - 4.9t) = 0$$

$$t = 0 \text{ or } t = \frac{10}{4.9} = 2.04 \text{ s}$$

The ball is at ground level when $t = 0$ and when $t = 2.04$, so the time of flight is $\frac{10}{4.9} = 2.04$ s.

(b) The range is the horizontal distance travelled by the ball. This can be found by substituting the time of flight into the expression for x.

$$x = 20 \cos 30° \times \frac{10}{4.9} = 35.3 \text{ m}$$

(c) The maximum height is attained when the vertical component of the velocity of the ball is zero.

$$0 = 20 \sin 30° - gt$$

$$0 = 10 - 9.8t$$

$$t = \frac{10}{9.8}$$

Using this value for t gives:

$$y = 10 \times \frac{10}{9.8} - 4.9\left(\frac{10}{9.8}\right)^2 = 5.10 \text{ m}$$

Worked example 7.2

A golf ball is hit from ground level, so that its initial velocity is 20 m s^{-1} at an angle of $40°$ above the horizontal.

(a) Find the time of flight of the ball.

(b) Find the range of the ball.

(c) Find the maximum height of the ball.

Solution

The position of the ball at time t is given by:

$$x = 20 \cos 40° \, t \quad \text{and} \quad y = 20 \sin 40° \, t - 4.9 \, t^2$$

(a) The ball is at ground level when $y = 0$, that is when

$$20 \sin 40° \, t - 4.9 \, t^2 = 0.$$

This equation can now be solved for t.

$$t(20 \sin 40° - 4.9 \, t) = 0$$

$$t = 0 \quad \text{or} \quad 20 \sin 40° - 4.9 \, t = 0$$

$$t = \frac{20 \sin 40°}{4.9} = 2.62 \text{ s}$$

(b) The range can be found by substituting the time of flight into the expression for x.

$$x = 20 \cos 40° \, t$$

$$= 20 \cos 40° \times \frac{20 \sin 40°}{4.9}$$

$$= 40.2 \text{ m}$$

(c) The ball is at its maximum height when the vertical component of its velocity is zero. The vertical component is given by $V \sin \theta - gt$. So in this case,

$$20 \sin 40° - 9.8 \, t = 0$$

$$t = \frac{20 \sin 40°}{9.8} = 1.31 \text{ s}$$

This value for t can now be substituted into the expression for y to give:

$$y = 20 \sin 40° \times \frac{20 \sin 40°}{9.8} - 4.9 \times \left(\frac{20 \sin 40°}{9.8}\right)^2$$

$$= 8.43 \text{ m}$$

So the maximum height of the ball is 8.43 m.

EXERCISE 7A

1 A rugby ball is kicked from ground level so that its initial velocity is 18 m s^{-1} and at an angle of 45° above the horizontal. Find:

   (a) the time of flight,

   (b) the range,

   (c) the maximum height of the ball.

2 David kicks a ball, from ground level, with a speed of 15 m s^{-1} at an angle of 30° to the horizontal. How far away from him does the ball land?

3 A bushbaby makes hops with a take-off speed of 6 m s^{-1} and at an angle of 30° to the horizontal. How far does it go in each hop?

4 A ball is thrown with initial speed 20 m s^{-1} at an angle of 60° to the horizontal. Assume that the ball is initially at ground level. How high does it rise? How far has it then travelled horizontally?

5 A stunt motorcyclist takes off at a speed of 35 m s^{-1} up a ramp of 30° to the horizontal to clear a river 50 m wide. Does the cyclist succeed in doing this? Does the motorcyclist have to worry about air resistance?

6 A ball is thrown with a speed of 12 m s^{-1} at an angle of 30° to the horizontal. It is initially at ground level.

   (a) Find the time of flight, the maximum height to which the ball rises and the range.

   (b) Find also the speed and direction of flight of the ball after 0.5 s and 1.0 s.

7 A golf ball is struck at a point O on the ground and moves with an initial velocity of 20 m s^{-1} at an angle of 53° to the horizontal. The ball subsequently lands at a point X which is on the same horizontal level as Q.

   (a) Show that the time taken by the ball to reach the point X is approximately 3.26 seconds.

   (b) Calculate the distance OX.

   (c) State:

      (i) the least speed of the ball during its flight from O to X,

      (ii) the direction of motion of the ball when this least speed occurs.   **[A]**

8 A golfer hits a ball, from ground level on a horizontal surface. The initial velocity of the ball is 21 m s^{-1} at an angle of 60° above the horizontal. Assume that the ball is a particle and that no resistance forces act on the ball.

   (a) Find the maximum height of the ball.

   (b) Find the range of the ball.

   (c) Find the speed of the ball at its maximum height.   [A]

9 Paul throws a ball from a point *O* with velocity 21 m s^{-1} at an angle of α to the horizontal, where $\sin \alpha = 0.7$. The ball subsequently moves freely under gravity in a vertical plane, as shown in the diagram.

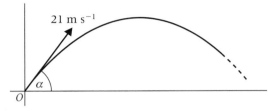

   (a) Show that the time taken for the ball to reach its greatest height above *O* is 1.5 seconds.

   (b) When the ball reaches its greatest height, it passes over a tree of vertical height 8 metres, as shown in the diagram.

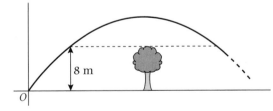

   (i) Find the vertical distance between the ball and the top of the tree at this time.

   (ii) Find the time between the ball leaving *O* and first reaching the horizontal level of the top of the tree. Give your answer to two decimal places.

   (iii) Find the length of time for which the ball is above the horizontal level of the top of the tree.   [A]

10 A particle *P* is projected at time $t = 0$ in a vertical plane from a point *O* with speed *u* at an angle α above the horizontal. Write down expressions for the horizontal and vertical components of:

   (a) the velocity of *P* at time *t*,

   (b) the displacement, at time *t*, of *P* from *O*.

(c) Given that the particle strikes the horizontal plane through O at time T show that

$$T = \frac{2u \sin \alpha}{g}$$

Find, in terms of g and T, the maximum height that P rises above the horizontal plane through O.   [A]

7.3 Including the height of release

A projectile will not always be launched from ground level.

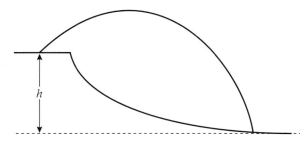

> The expression for y must be modified to take account of this height of release. If the height of release is h m, then
>
> $$y = V \sin \theta\, t - \frac{1}{2}gt^2 + h$$

The next example illustrates a case where the height of release must be included.

Worked example 7.3

A basket ball is thrown from a height of 1 m at a speed of 10 m s^{-1} and at an angle of 40° above the horizontal. The ball passes through the basket, which is at a height of 3 m above ground level. Find the horizontal distance of the basket from the point where the ball was thrown.

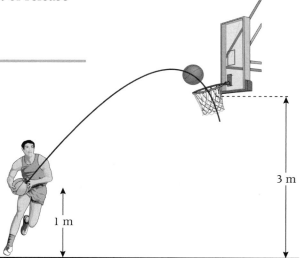

Solution

The position of the ball, at time t seconds, is given by:

$$x = V \cos \theta\, t \quad \text{and} \quad y = V \sin \theta\, t - \frac{1}{2}gt^2 + h$$

$$= 10 \cos 40°t \qquad\qquad = 10 \sin 40°t - 4.9t^2 + 1$$

The basket is at a height of 3 m, so when the ball enters the basket $y = 3$. This gives the quadratic equation:

$$3 = 10 \sin 40°t - 4.9t^2 + 1$$
$$4.9t^2 - 10 \sin 40°t + 2 = 0$$

Solving this quadratic equation gives the two times when the ball will be at a height of 3 m.

$$t = \frac{10 \sin 40° \pm \sqrt{100 \sin^2 40° - 4 \times 4.9 \times 2}}{2 \times 4.9}$$

$$= 0.5074 \text{ or } 0.8044 \text{ seconds (to four significant figures)}$$

As the ball moves up and then down, it will pass downwards through the basket at the later time. Taking this time and substituting into the expression for x gives:

$$x = 10 \cos 40° \times 0.8044$$

$$= 6.16 \text{ m (to three significant figures)}$$

So the basket is at a horizontal distance of 6.16 m from the point where the ball was thrown.

Working with the components of the velocity

In some problems the initial velocity of the projectile may not be known, or may be specified in terms of its horizontal and vertical components. In these cases the initial velocity can be expressed as

$$\mathbf{u} = U\mathbf{i} + V\mathbf{j}$$

so that the expressions for x and y become

$$x = Ut \quad \text{and} \quad y = Vt - \frac{1}{2}gt^2$$

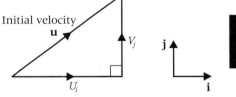

Worked example 7.4

An arrow is fired from a bow. It hits a target at the same level at a distance of 91 m from where it was fired and after 1.4 seconds.

(a) Find the horizontal and vertical components of the initial velocity.

(b) A cable passes over the area where the arrow was fired. It is at a height of 6 m and is at a horizontal distance of 20 m from where the arrow was fired. The arrow is fired at a height of 1 m, find the height of the cable above the arrow, when it passes beneath the cable.

Solution

The position of the arrow is given by:

$$x = Ut \quad \text{and} \quad y = Vt - 4.9t^2$$

where U and V are the horizontal and vertical components of the initial velocity, respectively.

(a) First use $x = 91$ and $t = 1.4$, to find U.

$$91 = U \times 1.4$$
$$U = 65 \text{ m s}^{-1}$$

Then use $y = 0$ and $t = 1.4$, to find V.

$$0 = V \times 1.4 - 4.9 \times 1.4^2$$
$$V = 4.9 \times 1.4 = 6.86 \text{ m s}^{-1}$$

(b) First revise the expression for y to take account of the height at which the arrow was fired. The revised expression is

$$y = Vt - 4.9t^2 + 1$$

The arrow passes under the cable when $x = 20$. This can be used to find the time when the arrow is directly below the cable.

$$20 = Ut$$
$$20 = 65t$$
$$t = \frac{20}{65} = \frac{4}{13} \text{ seconds}$$

This value of t can then be substituted into the revised expression for y to give the height of the arrow when it is below the cable.

$$y = 6.86 \times \frac{4}{13} - 4.9 \times \left(\frac{4}{13}\right)^2 + 1 = 2.65 \text{ m}$$

(to three significant figures)

So the distance between the arrow and the cable is given by

$$6 - 2.65 = 3.35 \text{ m}$$

The arrow passes 3.35 m below the cable.

Worked example 7.5

A particle is projected, so that the horizontal and vertical components of its initial velocity are U and $2U$, respectively. Find the range of this particle in terms of U and g.

Solution

The height, y, of the particle will be given by:

$$y = 2Ut - \frac{1}{2}gt^2.$$

The time of flight can be found by considering when y is zero, which gives the equation below.

$$0 = 2Ut - \frac{1}{2}gt^2$$
$$0 = t\left(2U - \frac{1}{2}gt\right)$$
$$t = 0 \quad \text{or} \quad t = \frac{4U}{g}$$

The horizontal distance travelled by the particle, x, is given by:

$$x = Ut$$

The time of flight can be substituted into this expression to give the range.

$$x = U \times \frac{4U}{g}$$

$$= \frac{4U^2}{g}$$

Worked example 7.6

A bullet is fired horizontally at a speed of 50 m s^{-1} from the top of a vertical cliff. The height of the cliff is 12 m. The bullet enters the sea at the point P.

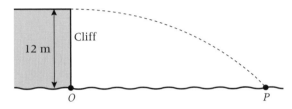

Find the distance OP.

Solution

The first step is to find the time when the bullet enters the sea at P. At time t the position of the bullet relative to O is given by:

$$x = 50t \quad y = 12 - 4.9t^2$$

When it hits the sea, $y = 0$, which gives

$$12 - 4.9t^2 = 0$$

$$t^2 = \frac{12}{4.9}$$

$$t = \sqrt{\frac{12}{4.9}} = 1.56 \text{ s}$$

Note that in this case we are only interested in the positive solution.

This value can then be substituted into the expression for x to find OP.

$$x = 50 \times \sqrt{\frac{12}{4.9}} = 78.2 \text{ m}$$

So the distance OP is 78.2 m.

EXERCISE 7B

1 An athlete launches a shot from a height of 2 m with an initial speed of 10 m s^{-1} and at an angle of 40° above the horizontal.

   (a) Find the range of the shot.

   (b) If the throw took place in an indoor arena, find the minimum height of the roof.

2 A javelin is modelled as a particle. Assume that only gravity acts on the javelin after it has left the thrower's hand. The initial velocity of the javelin is 20 m s^{-1} at an angle of 40° above the horizontal.

   (a) Find the range of the javelin on horizontal ground if the height of release is ignored.

   (b) The javelin is actually released at a height of 2 metres. Find the range of the javelin in this case.   [A]

3 The horizontal and vertical components of the initial velocity of a cricket ball are 20 m s^{-1} and 25 m s^{-1}. Assume that $g = 10$ m s^{-2}.

   (a) Find the range of the projectile on a horizontal surface.

   (b) If the ball is caught when it is at a height of 1.5 m and travelling downwards, find the time of flight and the horizontal distance travelled by the ball.

4 A stone is thrown up at an angle of 30° to the horizontal with a speed of 20 m s^{-1} from the edge of a cliff 15 m above sea level so that the stone lands in the sea. Find how long the stone is in the air and how far from the base of the cliff it lands.

5 A golf ball is hit from a position that is 4 metres higher than the horizontal area where the ball lands. The initial velocity of the ball is 30 m s^{-1} at an angle of 60° above the horizontal.

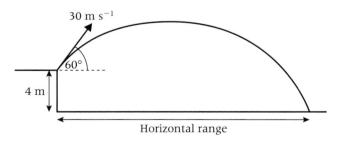

   (a) Show that the maximum height of the ball, above the landing area, is approximately 38.4 metres.

   (b) Show that the ball hits the ground approximately 5.45 seconds after it has been hit.

   (c) Hence calculate the horizontal range of the ball to the nearest metre.   [A]

6

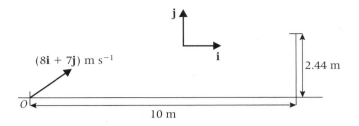

During a game of football, Tom kicks the ball towards the goal. He kicks the ball with velocity $(8\mathbf{i} + 7\mathbf{j})$ m s^{-1}, from a point O which is 10 metres from the goal. The height of the top of the goal is 2.44 metres.

The unit vectors $\mathbf{i}$ and $\mathbf{j}$ are horizontal and vertically upwards, respectively, as shown in the diagram.

(a) Show that the time the ball takes to reach the goal is 1.25 seconds.

(b) Determine whether the ball passes under or over the top of the goal.

(c) Find the speed of the ball when it reaches the goal, giving your answer to two significant figures.   [A]

7 Karen is standing 4 m away from a wall which is 2.5 m high. She throws a ball at 10 m s^{-1} at an angle of 40° to the horizontal from a height of 1 m above the ground. Will the ball pass over the wall?

8 A stone is thrown with speed 10 m s^{-1} at an angle of projection of 30° from the top of a cliff and hits the sea 2.5 s later.

(a) How high is the cliff?

(b) How far from the base of the cliff does the stone hit the water?

9 A bowler releases a cricket ball from a height of 2.25 m above the ground so that initially its velocity is horizontal. Find the speed of delivery if it is to hit the ground a horizontal distance of 16 m from the point of release.

10 A tennis ball is hit so that it initially moves horizontally at V m s^{-1} from a point at a height of 2 m. The net which has a height of 0.9 m is at a horizontal distance of 12 m from the point where the ball was hit.

(a) Find the minimum value of V for which the ball clears the net.

(b) If $V = 30$ m s^{-1} describe what happens to the ball when it has travelled 12 m horizontally.

7

11 A ball is thrown so that it passes through the centre of a basket ball hoop, as shown in the diagram.

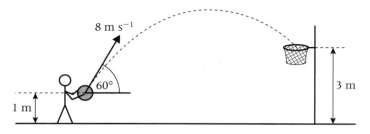

The ball is thrown from a height of 1 metre and the hoop is at a height of 3 metres above the ground. The initial velocity of the ball is 8 m s^{-1} at an angle of 60° above the horizontal.

(a) Find the maximum height of the ball above the ground.

(b) Find the time that it takes for the ball to reach the centre of the hoop.

(c) Find the horizontal distance from the initial position of the ball to the centre of the hoop. [A]

12 A tennis player plays a ball with speed 20 m s^{-1} horizontally straight down the court from the backline.

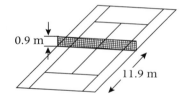

(a) What is the least height at which she can play the ball to clear the net?

(b) How far behind the net does the ball land when it is played at this height?

13 A golfer hits a golf ball so that it moves with an initial speed of 40 m s^{-1} at an angle of 20° above the horizontal.

(a) State two essential assumptions that you should make if you are to estimate the horizontal distance between the point where the ball was hit and the point where it hits the ground for the first time.

(b) Find this distance to the nearest metre.

(c) The ball is actually hit on a raised area that is 5 m higher than the ground where the ball lands for the first time. By how much would this increase the answer that you obtained in (b)?

14 Take $g = 10 \text{ m s}^{-2}$ in this question.

When a tennis ball is served, it is hit from the baseline A. It must pass over the net, at B, and land **between** the net and service line on the other side, at C. The diagram shows the positions of these lines and the height of the net.

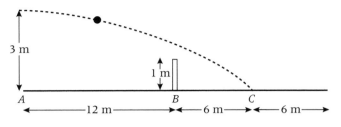

A tennis ball is served so that it initially moves horizontally at a speed of $v \text{ m s}^{-1}$ from a point 3 m above the baseline A. Assume that there are no resistance forces and that the ball moves in a vertical plane at right angles to the net.

(a) (i) Show that v must be greater than $\sqrt{360} \text{ m s}^{-1}$ if the ball is to go over the net.

(ii) What is the maximum value of v if the ball is to land in the area between B and C?

(b) A tennis player hits a tennis ball so that is initially moves at a speed of 30 m s^{-1} and at an angle of $5°$ below the horizontal. Is this serve successful? **[A]**

15

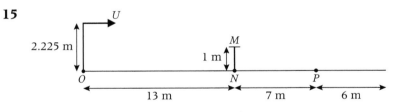

The diagram shows a cross-section of a court where Oliver is practising tennis shots. The court is of length 26 metres, and the net, situated at N, the centre of the court, is of height 1 metre. Oliver stands at the point O at one end of the court and serves the ball from a height of 2.225 metres above O with a horizontal velocity $U \text{ m s}^{-1}$.

(a) When $U = u_1$, the ball strikes point M, at the top of the net.

(i) By considering the vertical component of the motion of the ball, show that the time taken for the ball to reach M is 0.5 seconds.

(ii) Hence, find the value of u_1.

(b) When $U = u_2$, where $u_2 > u_1$, the ball first hits the court between the points N and P, where $NP = 7$ metres. Show that $u_2 < 30 \text{ m s}^{-1}$. **[A]**

7

16 George throws a ball from a point O, with velocity $\begin{bmatrix} 3U \\ 4U \end{bmatrix}$. The ball subsequently lands at a point H, which is at the same horizontal level as O, as shown in the diagram.

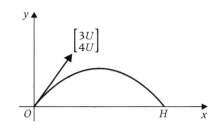

(a) Show that the time taken by the ball to travel from O to H is $\dfrac{8U}{g}$.

(b) Find, in terms of g and U, the distance OH.

(c) Find, in terms of U, the initial speed of the ball.

(d) Find, in terms of g and U, the two times during the flight from O to H when the ball is moving with speed $\sqrt{18}U$. [A]

17 A stuntman in a film drives a car off the top of a vertical cliff. The top of the cliff is 25 metres above the level of the sea. When it leaves the cliff the car is travelling at 40 m s^{-1} at an angle of 5° above the horizontal. The diagram shows the cliff and the initial velocity of the car.

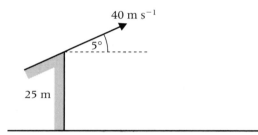

Model the car as a particle and assume that, while it is in the air, it moves under the influence of gravity alone.

(a) Show that the car hits the sea approximately 2.64 seconds after it leaves the top of the cliff.

(b) Find the horizontal distance of the car from the cliff when the car hits the sea.

(c) Find the speed of the car when it hits the sea. [A]

Key point summary

Formulae to learn:

$$x = V \cos \theta \, t$$

$$y = V \sin \theta \, t - \frac{1}{2}gt^2$$

- A projectile moves only under the influence of gravity *p110* and so it will have an acceleration of g m s^{-2}.

- The constant acceleration equations can be used to *p111* formulate expressions for velocity and position of a projectile at time t.

- The horizontal component of velocity of a projectile *p111* can be written as $x = V \cos \theta \, t$ and the vertical component of velocity can be written as $y = V \sin \theta \, t - \frac{1}{2}gt^2$.

- The height of release can be calculated by modifying *p116* the expression for y.

- Range will vary with the angle of projection. *p111*

Test yourself	What to review	**7**

1 A particle is projected from ground level on a horizontal *Section 7.?*
plane, with speed 30 m s^{-1} and at an angle of 60° above the horizontal.
  (a) Find the range of the particle.
  (b) Find the maximum height of the particle.

2 A bullet is fired horizontally at a speed of 100 m s^{-1} from a *Section 7.?*
height of 3 m.
  (a) Find the time when the bullet hits the ground.
  (b) Find the horizontal distance travelled by the bullet.

3 A ball is thrown from a height of 1.5 m at a speed of 12 m s^{-1} *Section 7.?*
and at an angle of 20° above the horizontal. It hits the ground
for the first time T seconds after it was thrown.
  (a) Find T.
  (b) Find the horizontal distance travelled by the ball.
  (c) Find the speed of the ball when it hits the ground.

Test yourself ANSWERS

3 (a) 1.11 s; (b) 12.5 m; (c) 13.2 m s^{-1}.

2 (a) 0.782 s; (b) 78.2 m.

1 (a) 79.5 m; (b) 34.4 m.

CHAPTER 8

Momentum

Learning objectives

After studying this chapter you should be able to:
- define the momentum of an object
- understand what happens to momentum of a particle during collision
- use the principle of conservation of momentum to calculate momentum after collision, in both one and two dimensions.

8.1 Introduction

In this chapter we will consider a quantity related to the motion of a body called its momentum. The momentum is defined as the product of the mass and the velocity of the body that is moving. The main focus of the chapter will be to consider what happens to the momentum of bodies that are involved in collisions.

8.2 Momentum

> The momentum of an object is defined as mv where m is the mass and v is the velocity.

Worked example 8.1

A car has mass 1200 kg and is travelling at 20 m s^{-1}. A lorry has mass 3500 kg and is travelling in the opposite direction to the car at 12 m s^{-1}.

Calculate the momentum of the car and the momentum of the lorry.

Solution

Defining the direction in which the car is travelling as positive gives:

Momentum of car $= 1200 \times 20 = 24\,000 \text{ kg m s}^{-1}$

Momentum of lorry $= 3500 \times (-12) = -42\,000 \text{ kg m s}^{-1}$

Note that the units for momentum are kg m s^{-1} or alternatively N s.

Worked example 8.2

A ball has mass 200 grams. When it is hit by a bat its velocity changes from $\begin{bmatrix} 4 \\ 10 \end{bmatrix}$ m s^{-1} to $\begin{bmatrix} -16 \\ -20 \end{bmatrix}$ m s^{-1}. Find the change in the momentum of the ball.

Solution

$$\text{Momentum before} = 0.2 \times \begin{bmatrix} 4 \\ 10 \end{bmatrix}$$

$$= \begin{bmatrix} 0.8 \\ 2 \end{bmatrix}$$

$$\text{Momentum after} = 0.2 \times \begin{bmatrix} -16 \\ -20 \end{bmatrix}$$

$$= \begin{bmatrix} -3.2 \\ -4 \end{bmatrix}$$

$$\text{Change in momentum} = \begin{bmatrix} -3.2 \\ -4 \end{bmatrix} - \begin{bmatrix} 0.8 \\ 2 \end{bmatrix}$$

$$= \begin{bmatrix} -4 \\ -6 \end{bmatrix} \text{N s}$$

8.3 **Conservation of momentum**

We will now consider what happens to the momentum of a particle that is involved in a collision.

Consider two particles that are involved in a collision. The diagrams below show the velocities of the particles before and after the collision.

While the particles are in contact they exert equal, but opposite forces on each other. If the magnitude of each of these forces is assumed to be a constant F, then we can find a relationship between the accelerations of the particles during the collision.

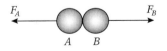

$$-F_A = F_B$$
$$-m_A a_a = m_B a_B$$

As the particles are in contact for the same period of time, t, we can also state that:

$$-m_A a_A t = m_B a_B t \quad [1]$$

The constant acceleration equation $v = u + at$ can be used to eliminate t and the accelerations from our equation. Using $a_A t = v_A - u_A$ and $a_B t = v_B - u_B$. Then substituting into expression [1] gives:

$$m_A a_A t = m_B a_B t$$
$$-m_A(v_A - u_A) = m_B(v_B - u_B)$$
$$m_A u_A + m_B u_B = m_A v_A + m_B v_B$$

or

total momentum before collision
= total momentum after collision

This is called the principle of conservation of momentum. It can be applied to any collision provided that no external forces act during the collision.

Worked example 8.3

A toy train, of mass 200 grams, is moving along a straight track at 1.8 m s^{-1}, when it collides with a stationary carriage of mass 300 grams. During the collision the train and carriage are coupled together. Find the speed of the train and carriage after the collision.

Solution

The diagram shows the velocities of the trucks before and after the collision.

After the collision both the train and the carriage have the same speed, v. So applying the principle of conservation of momentum gives:

$$0.2 \times 1.8 + 0.3 \times 0 = (0.2 + 0.3)v$$
$$0.36 = 0.5v$$
$$v = 0.72 \text{ m s}^{-1}$$

Worked example 8.4

Two particles travel towards each other along a straight line. One has mass 3 kg and speed 4 m s^{-1}. The other has mass 5 kg and speed 2 m s^{-1}. When they collide the 3 kg mass is brought to rest. What happens to the 5 kg mass?

Solution

The diagrams show the velocities of the particles before and after the collision.

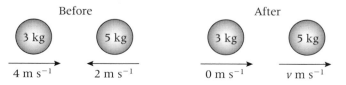

Applying the principle of conservation of momentum gives:

$$3 \times 4 + 5 \times (-2) = 3 \times 0 + 5 \times v$$
$$2 = 5v$$
$$v = 0.4 \text{ m s}^{-1}$$

Worked example 8.5

A person, of mass 60 kg, stands on a trolley of mass 20 kg, that is free to move on a horizontal surface. On the trolley there is also a large medicine ball of mass 5 kg. Initially the trolley is at rest. The person then throws the ball off the trolley so that its initial velocity is 4 m s^{-1} horizontally. Find the final speed of the trolley.

Solution

The diagram shows the velocities before and after the throw.

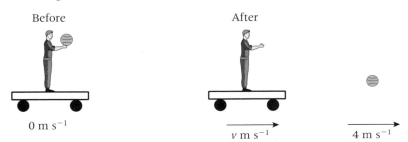

Using the principle of conservation of momentum gives:

$$85 \times 0 = 80v + 5 \times 4$$
$$v = -\frac{20}{80}$$
$$= -0.25 \text{ m s}^{-1}$$

So the trolley moves at 0.25 m s^{-1}, after the throw. Note that the trolley travels in the opposite direction to the ball.

EXERCISE 8A

1 Calculate the momentum of:
  (a) a train, of mass 120 tonnes, travelling at 40 m s^{-1},
  (b) a table tennis ball, of mass 3 g, travelling at 4 m s^{-1},
  (c) a car, of mass 1200 kg, travelling at 36 kph.

2 In the diagram A and B have masses 2 kg and 3 kg and move with initial velocities as shown. The collision reduces the velocity of A to 2 m s^{-1}. Find the velocity of B after the collision.

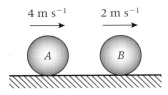

3 The two bodies in the diagram have masses 3 kg and 5 kg, respectively. They are travelling with speeds $4\,\text{m s}^{-1}$ and $2\,\text{m s}^{-1}$. The bodies coalesce on impact. Find the speed of this body.

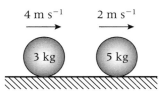

4 A particle A of mass 250 grams collides with a particle B of mass 150 grams. Initially A has velocity $7\,\text{m s}^{-1}$ and B is at rest. After the collision, the velocity of B is $5\,\text{m s}^{-1}$. Calculate the velocity of A after the impact.

5 Two railway trucks, each of mass 8 tonnes, are travelling in the same direction and along the same tracks with velocities $3\,\text{m s}^{-1}$ and $1\,\text{m s}^{-1}$, respectively. When the trucks collide they couple together. Calculate the velocity of the coupled trucks.

6 A child of mass 40 kg jumps onto a stationary skateboard of mass 4 kg. If the child was moving horizontally at $3.3\,\text{m s}^{-1}$ just before he lands on the skateboard, find the speed at which they move.

7 A tow truck of mass 3 tonnes is attached to a car of mass 1.2 tonnes by a rope. The truck is moving at a constant $3\,\text{m s}^{-1}$ when the tow rope becomes taut and the car begins to move. Assume that both vehicles move at the same speed once the rope is taut, and find this speed.

8 A red snooker ball that was at rest is hit directly by a white ball moving at $0.8\,\text{m s}^{-1}$. After the collision the red ball moves at $0.75\,\text{m s}^{-1}$. Assume that all the motion takes place on a straight line. Find the speed of the white ball after the collision.

9 Two identical uniform smooth spheres, A and B, each of mass m, moving in opposite directions with speeds u and $3u$, respectively, collide directly. The sphere B is brought to rest by the collision. Find the speed of A after the collision and state the direction in which it moves.

10 A car of mass 1.2 tonnes collides with a stationary van of mass 2.4 tonnes. After the collision the two vehicles become entangled and skid 15 m before stopping. Police accident investigators estimate that the magnitude of the friction force during the skid was 2880 N. Assume the road is horizontal and that all the motion takes place in a straight line.

   (a) Find the speed of the vehicles just after the collision.

   (b) Find the speed of the car before the collision.   [A]

11 A train of total mass 110 tonnes and velocity 80 km h^{-1} crashes into a stationary locomotive of mass 70 tonnes.

   (a) Calculate the velocity of the combined system immediately after impact.

   The trains plough on for a further 40 m.

   (b) Calculate the average deceleration and the resistance to motion.

12 A pile-driver consists of a pile of mass 200 kg and a driver of mass 40 kg. The driver drops on the pile with velocity 6 m s^{-1} and sticks to the top of the pile.

   (a) Calculate the velocity of the pile immediately after impact.

   Resistances to motion of the pile amount to 1400 N.

   (b) Calculate the distance penetrated by the pile.

13 Two uniform smooth spheres, *A* of mass 0.03 kg and *B* of mass 0.1 kg, have equal radii and are moving directly towards each other with speeds of 7 m s^{-1} and 4 m s^{-1}, respectively. The spheres collide directly and *B* is reduced to rest by the impact. Find the speed of *A* after the impact.   [A]

14 Two cars are initially 36 m apart travelling in the same direction along a straight, horizontal road. The car in front is initially travelling at 10 m s^{-1}, but decelerating at 2 m s^{-2}. The other car travels at a constant 15 m s^{-1}.

   (a) Model the cars as particles. By finding the distance travelled by each car after *t* seconds, show that the distance between the two cars is $36 - 5t - t^2$ metres. Find when they would collide if neither car takes avoiding action.

   (b) Would it be necessary to revise your answers to **(a)** if the cars were not modelled as particles? Give reasons to support your answer.

   The mass of the front car is 1500 kg and the mass of the other car is 1000 kg.

   (c) The cars do collide and after the collision the two cars move together. Find the speed of the cars just after the collision.   [A]

15 A stone, *A*, of mass 0.05 kg, is sliding in a straight line with speed 3 m s^{-1} across a smooth frozen pond when it collides directly with a stationary stone, *B*, of mass 0.2 kg. After the collision, *A* and *B* move directly **away** from each other, each with speed *v* m s^{-1}.

   Find the value of *v*.   [A]

8

16 Two particles, *A* and *B*, of masses $2m$ kg and m kg, respectively, are moving directly **towards** each other on a smooth horizontal surface. The speeds of *A* and *B* are 2 m s^{-1} and 6 m s^{-1}, respectively.

The particles *A* and *B* collide and subsequently move directly **away** from each other with speeds $3V$ m s^{-1} and V m s^{-1}, respectively.

Find the value of *V*. [A]

17 A particle, *P*, has mass 5 kg. It is moving along a straight line with speed 4 m s^{-1}, when it collides directly with another particle, *Q*, which is at rest. The mass of *Q* is m kg.

5 kg   *m* kg

P   *Q*

4 m s^{-1}

After the collision *P* moves with a speed of 1.2 m s^{-1} and *Q* moves with a speed of 1.4 m s^{-1}.

(a) If *P* and *Q* both move in the same direction after the collision, show that $m = 10$.

(b) If *P* and *Q* move in opposite directions after the collision, find *m*. [A]

18 A bullet, of mass 50 grams, is fired from a gun, of mass 5 kg. The bullet hits a wooden block, of mass 3 kg, that is initially at rest on a rough, horizontal surface. The bullet becomes embedded in the block. The combined block and bullet slide 0.5 m before coming to rest. The coefficient of friction between the block and the surface is 0.8. Assume that the bullet always travels horizontally.

(a) Show that the speed of the bullet before it hits the block is 170.8 m s^{-1}.

(b) Find the speed at which the gun recoils after firing. [A]

19 A sledge, of mass 10 kg, is at rest on an icy, horizontal surface. A child, of mass 40 kg, is standing on the sledge.

The child jumps off the sledge. Initially the child travels horizontally at 2 m s^{-1} and the sledge begins to slide in the opposite direction to the child.

(a) Assuming that momentum is conserved, find the speed of the sledge, just after the child has jumped off it.

(b) The coefficient of friction between the sledge and the ice is 0.2.
    (i) Find the magnitude of the friction force acting on the sledge while it is moving.
    (ii) Find the distance that the sledge slides before it comes to rest. [A]

20 After a collision, a car and a van, of combined mass 3000 kg, slide together along a straight horizontal road. The coefficient of friction between the road and the tyres of the vehicles as they slide is 0.7.

   (a) Model the car and the van as a single particle.

      (i) Show that the magnitude of the frictional force acting is 20 580 N.

      (ii) Find the acceleration of the car and van after the collision.

      (iii) The car and the van slide together for a distance of 5 m before coming to rest.

         Using the result from **(a)(ii)**, show that just after the collision the car and van were moving at 8.28 m s^{-1}, to three significant figures.

   (b) The mass of the car is 1200 kg and the mass of the van is 1800 kg. Before the collision, the van was stationary. Find the speed of the car just before the collision.   [A]

21 Three supermarket trolleys, each of mass 20 kg, are placed on a straight line. As these trolleys move they are not subject to any resistance to motion. The first one is set in motion so that it moves at a speed of 4.5 m s^{-1} towards the second trolley. It then collides with the second trolley. After this collision the two trolleys continue to move together along the straight line at constant speed until they collide with the third trolley. After this collision all three trolleys move together along the straight line.

   (a) Show that the speed of the two moving trolleys after the first collision is 2.25 m s^{-1}.

   (b) Find the speed of the trolleys after the second collision.

   (c) After the second collision the combined trolleys are subject to a single, horizontal resistance force of magnitude 30 N.

      (i) Calculate the acceleration of the trolleys while this force acts.

      (ii) Find the distance that the trolleys move after the second collision, before they come to rest.   [A]

22 A test is carried out on a rocket, of mass 200 kg, which is fired horizontally at a speed of 50 m s^{-1}. The rocket experiences a constant air resistance force of 1568 N. It travels a distance of 108 m before it hits a stationary tank, of mass 4800 kg. Assume that the rocket always travels horizontally.

   (a) Find the speed of the rocket when it has travelled a distance of 108 m.

8

When the rocket hits the tank it becomes lodged in it.

(b) Find the speed of the tank and the rocket just after the collision.

The tank then slides until it comes to rest. The coefficient of friction between the tank and the ground is 0.6. Neglect any air resistance when considering the motion of the tank and rocket together.

(c) Find the distance that the tank slides.   [A]

8.4 Conservation of momentum in two dimensions

In the previous section you will have applied the principle of conservation of momentum in one dimension. In this section we will apply this principle in two dimensions. In order to do this we will need to work with vectors.

> The principle of the conservation of momentum, in vector form, is:
>
> $$m_A\mathbf{u}_A + m_B\mathbf{u}_B = m_A\mathbf{v}_A + m_B\mathbf{v}_B$$

Worked example 8.6

Particles *A* and *B* collide and coalesce.
Particle *A* has mass 6 kg and had velocity $(3\mathbf{i} + 5\mathbf{j})$ m s^{-1} before the collision.
Particle *B* has mass 4 kg and had velocity $(4\mathbf{i} - 9\mathbf{j})$ m s^{-1} before the collision.
Find the velocity of the combined particle after the collision.

Solution

As the particles coalesce there will be a single particle of mass 10 kg moving with velocity **v** after the collision.

Applying the principle of conservation of momentum gives:

$$6(3\mathbf{i} + 5\mathbf{j}) + 4(4\mathbf{i} - 9\mathbf{j}) = 10\mathbf{v}$$
$$18\mathbf{i} + 30\mathbf{j} + 16\mathbf{i} - 36\mathbf{j} = 10\mathbf{v}$$
$$34\mathbf{i} - 6\mathbf{j} = 10\mathbf{v}$$

Then dividing by 10 gives:

$$\mathbf{v} = 3.4\mathbf{i} - 0.6\mathbf{j}$$

Alternatively column vectors could be used to solve this problem, as shown below:

$$6\begin{bmatrix} 3 \\ 5 \end{bmatrix} + 4\begin{bmatrix} 4 \\ -9 \end{bmatrix} = 10\mathbf{v}$$

$$\begin{bmatrix} 18 \\ 30 \end{bmatrix} + \begin{bmatrix} 16 \\ -36 \end{bmatrix} = 10\mathbf{v}$$

$$\begin{bmatrix} 34 \\ -6 \end{bmatrix} = 10\mathbf{v}$$

$$\mathbf{v} = \begin{bmatrix} 3.4 \\ -0.6 \end{bmatrix}$$

Worked example 8.7

Two roads are perpendicular and meet at a junction.
Car *A*, of mass 1400 kg, travels along one road at 20 m s^{-1}.
A second car, *B*, of mass 1100 kg, travels along the other road at
16 m s^{-1}.
At the junction the cars collide and then move together.

(a) Write down the initial velocities of the cars before the collision.

(b) Find the velocity of the cars after the collision.

Solution

The diagram shows the velocities of the two cars and the unit
vectors **i** and **j**.

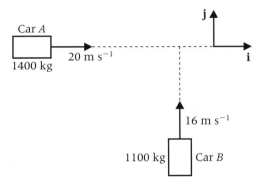

(a) The initial velocities can now be written as

$$\mathbf{u}_A = 20\mathbf{i} \quad \text{and} \quad \mathbf{u}_B = 16\mathbf{j}.$$

(b) Applying the principle of conservation of momentum gives

$$1400 \times 20\mathbf{i} + 1100 \times 16\mathbf{j} = (1400 + 1100)\mathbf{v}$$

$$28\,000\mathbf{i} + 17\,600\mathbf{j} = 2500\mathbf{v}$$

$$\mathbf{v} = \frac{28\,000}{2500}\mathbf{i} + \frac{17\,600}{2500}\mathbf{j} = 11.2\mathbf{i} + 7.04\mathbf{j}$$

Worked example 8.8

Two particles, *A* of mass 3 kg and *B* of mass 5 kg collide.
Before the collision the velocity of *A* is $(3\mathbf{i} - 2\mathbf{j})$ m s^{-1} and the
velocity of *B* is $(5\mathbf{i} + 3\mathbf{j})$ m s^{-1}. After the collision the velocity of *A*
is $(2\mathbf{i} + \mathbf{j})$ m s^{-1}. Find the velocity of *B* after the collision.

Solution

In this case, we have $\mathbf{u}_A = 3\mathbf{i} - 2\mathbf{j}$, $\mathbf{u}_B = 5\mathbf{i} + 3\mathbf{j}$ and $\mathbf{v}_A = 2\mathbf{i} + \mathbf{j}$.

Applying the principle of conservation of momentum and simplifying gives

$$3(3\mathbf{i} - 2\mathbf{j}) + 5(5\mathbf{i} + 3\mathbf{j}) = 3(2\mathbf{i} + \mathbf{j}) + 5\mathbf{v}_B$$
$$9\mathbf{i} - 6\mathbf{j} + 25\mathbf{i} + 15\mathbf{j} = 6\mathbf{i} + 3\mathbf{j} + 5\mathbf{v}_B$$
$$34\mathbf{i} + 9\mathbf{j} = 6\mathbf{i} + 3\mathbf{j} + 5\mathbf{v}_B$$
$$28\mathbf{i} + 6\mathbf{j} = 5\mathbf{v}_B$$

Now dividing by 5 gives the velocity of B after the collision:

$$\mathbf{v}_B = \frac{28}{5}\mathbf{i} + \frac{6}{5}\mathbf{j} = 5.6\mathbf{i} + 1.2\mathbf{j}$$

Worked example 8.9

Two particles, P and Q, collide, and after the collision they move together in the direction of the unit vector $\mathbf{i}$. Before the collision the velocity of A is $(4\mathbf{i} + 5\mathbf{j})$ m s^{-1} and the velocity of B is $(-2\mathbf{i} - 4\mathbf{j})$ m s^{-1}. The mass of A is 2 kg and the mass of B is m kg.

Find the value of m and the velocity of the combined particle after the collision.

Solution

In this case $\mathbf{u}_A = 4\mathbf{i} + 5\mathbf{j}$ and $\mathbf{u}_B = -2\mathbf{i} - 4\mathbf{j}$. After the collision the combined particle will have velocity $\mathbf{v} = k\mathbf{i}$, where k is a constant.

Applying the principle of conservation of momentum gives

$$2(4\mathbf{i} + 5\mathbf{j}) + m(-2\mathbf{i} - 4\mathbf{j}) = (m + 2)(k\mathbf{i})$$
$$8\mathbf{i} + 10\mathbf{j} - 2m\mathbf{i} - 4m\mathbf{j} = (m + 2)k\mathbf{i}$$
$$(8 - 2m)\mathbf{i} + (10 - 4m)\mathbf{j} = (m + 2)k\mathbf{i}$$

The $\mathbf{j}$ component must be zero on both sides of the equation, so

$$10 - 4m = 0$$
$$m = \frac{10}{4} = 2.5$$

This value of m can now be substituted into the conservation of momentum equation, which gives

$$(8 - 2 \times 2.5)\mathbf{i} + (10 - 4 \times 2.5)\mathbf{j} = (2.5 + 2)k\mathbf{i}$$
$$3\mathbf{i} = 4.5k\mathbf{i}$$

Hence,

$$3 = 4.5k$$

$$k = \frac{3}{4.5} = \frac{2}{3}$$

After the collision the velocity of the particles is $\left(\frac{2}{3}\mathbf{i}\right)$ m s^{-1}.

EXERCISE 8B

1 Two particles, A and B, collide. After the collision they move together as a single particle. The mass of A is 12 kg and the mass of B is 8 kg. Before the collision, the velocity of A is $(4\mathbf{i} - 6\mathbf{j})$ m s^{-1} and the velocity of B is $(3\mathbf{i} + 2\mathbf{j})$ m s^{-1}. Find the velocity and the speed of the combined particle after the collision.

2 Particle A has mass 2 kg and velocity $\begin{bmatrix} 6 \\ -2 \end{bmatrix}$ m s^{-1} when it collides with particle B, which has mass 3 kg and velocity $\begin{bmatrix} 10 \\ 8 \end{bmatrix}$ m s^{-1}. After the collision the particles move together. Find the velocity of the combined particles after the collision.

3 Particle A has mass 5 kg and velocity $\begin{bmatrix} 3 \\ 7 \end{bmatrix}$ m s^{-1} when it collides with particle B, which has mass 4 kg. After the collision the particles move together with velocity $\begin{bmatrix} 1 \\ 4 \end{bmatrix}$ m s^{-1}. Find the velocity of B before the collision.

4 Two particles both have mass 5 kg and are moving with velocities $(9\mathbf{i} - 6\mathbf{j})$ m s^{-1} and $(6\mathbf{i} + 6\mathbf{j})$ m s^{-1} when they collide and coalesce. Describe the direction in which they move after the collision and their speed.

5 A particle has mass 2 kg and velocity $(7\mathbf{i} + 2\mathbf{j})$ m s^{-1} when it collides with a particle of mass m, which is moving parallel to the unit vector $\mathbf{j}$. After the collision the two particles move together with velocity $(2\mathbf{i} + \mathbf{j})$ m s^{-1}.

(a) Find m.

(b) Find the velocity of B before the collision.

6 Two particles, A and B, collide. The table gives the masses and velocities of the particles, where α and β are constants.

Particle	Mass	Velocity before collision	Velocity after collision
A	4 kg	$(3\mathbf{i} + 5\mathbf{j})$ m s^{-1}	$(\mathbf{i} + \alpha\mathbf{j})$ m s^{-1}
B	6 kg	$(-2\mathbf{i} + 6\mathbf{j})$ m s^{-1}	$(\beta\mathbf{i} + 2\mathbf{j})$ m s^{-1}

Find α and β.

8

7 When two particles collide they are both brought to rest. One particle has mass 5 kg and had velocity $\begin{bmatrix} 8 \\ -9 \end{bmatrix}$ m s^{-1} before the collision. The other particle has mass 2 kg. Find the velocity of this particle before the collision.

8 Two particles, *A* and *B*, collide. The table gives the masses and velocities of the particles.

Particle	Mass	Velocity before collision	Velocity after collision
A	1 kg	$(8\mathbf{i} + 15\mathbf{j})$ m s^{-1}	$(\mathbf{i} + 6\mathbf{j})$ m s^{-1}
B	4 kg	$(-\mathbf{i} - 2\mathbf{j})$ m s^{-1}	v m s^{-1}

Find **v**.

9 A sphere has mass 7 kg and velocity $(5\mathbf{i} - 8\mathbf{j})$ m s^{-1}. It collides with a stationary sphere of mass 5 kg. After the collision the 7 kg sphere has velocity $(2\mathbf{i} - 3\mathbf{j})$ m s^{-1}. Find the velocity of the other sphere after the collision.

10 Two particles, *A* of mass 7 kg and *B* of mass 9 kg, collide and coalesce. Before the collision the velocity of *A* was **v** and the velocity of *B* was 2**v**. After the collision their velocity is $\begin{bmatrix} 3 \\ 2 \end{bmatrix}$ m s^{-1}. Find **v**.

11 A particle *A*, of mass 0.3 kg, is moving with velocity $\begin{bmatrix} 7 \\ 4 \end{bmatrix}$ m s^{-1} when it collides with a stationary particle, *B*, of mass 0.5 kg. Immediately after the collision, *B* moves with velocity $\begin{bmatrix} 6 \\ 0 \end{bmatrix}$ m s^{-1}.

(a) Find the velocity of *A* immediately after the collision.

(b) Find the speed of *A* immediately after the collision.

(c) State which of *A* and *B* moves faster after the collision. [A]

12 A particle, *P*, moves with constant velocity.

(a) The particle passes through the points with position vectors $\begin{bmatrix} -1 \\ 2 \end{bmatrix}$ and $\begin{bmatrix} 2 \\ 8 \end{bmatrix}$ at times $t = 0$ and $t = 3$, respectively. The units of distance and time are metres and seconds, respectively. Show that the velocity of *P* is $\begin{bmatrix} 1 \\ 2 \end{bmatrix}$ m s^{-1}.

(b) When $t = 3$, the particle P, of mass 0.2 kg, collides with a particle Q. The particle Q has mass 0.1 kg and, immediately before the collision, it is moving with velocity $\begin{bmatrix} -5 \\ 5 \end{bmatrix}$ m s^{-1}. As a result of the collision, P and Q coalesce into a single particle, R. Find the velocity of R immediately after the collision.

(c) Assuming that the velocity of R remains constant, find the position vector of R three seconds after the collision. [A]

Key point summary

1 Momentum is the product of mass and velocity, *p126*

   that is momentum $= mv$

2 Momentum is conserved in collisions, when no *p127*
   external forces act.

3 Conservation of momentum *p127*

   $m_A u_A + m_B u_B = m_A v_A + m_B v_B$

4 The principle of the conservation of momentum, in *p134*
   vector form, is:

   $m_A \mathbf{u}_A + m_B \mathbf{u}_B = m_A \mathbf{v}_A + m_B \mathbf{v}_B$

8

Test yourself	What to review
1 A car, of mass 1000 kg, is travelling at 20 m s^{-1}, when it drives into a truck, of mass 4000 kg, which was moving at 10 m s^{-1} in the same direction. After the collision the two vehicles move together. Find the speed of the vehicles after the collision.	*Section 8.3*
2 A child of mass 40 kg stands on a skate board, of mass 3 kg. Initially both are at rest. The boy jumps off so that he travels horizontally at 1.2 m s^{-1}. Find the speed of the skateboard.	*Section 8.3*
3 Two particles are travelling towards each other when they collide. One has mass 2 kg and was travelling at 5 m s^{-1} before the collision and the other has mass 3 kg and a velocity of 6 m s^{-1} before the collision. The 2 kg mass reverses direction and moves at 2.5 m s^{-1} after the collision. Describe how the 3 kg mass moves after the collision.	*Section 8.3*

Test yourself (continued)	What to review

4 Two particles, *A* and *B*, have velocities $\begin{bmatrix} 4 \\ 7 \end{bmatrix}$ m s^{-1} and

Section 8.?

$\begin{bmatrix} -2 \\ 6 \end{bmatrix}$ m s^{-1}, respectively, when they collide. The mass of *A* is 4 kg and the mass of *B* is 6 kg.

  (a) If the particles coalesce, find their velocity after the collision.

  (b) If the velocity of *B* is $\begin{bmatrix} 0.5 \\ 4 \end{bmatrix}$ m s^{-1} after the collision, find the velocity of *A*.

5 The mass of particle *A* is 3 kg and it moves with velocity

Section 8.?

$\begin{bmatrix} 2 \\ 3 \end{bmatrix}$ m s^{-1}. The mass of particle *B* is *m* kg and it moves with

velocity $\begin{bmatrix} k \\ 11 \end{bmatrix}$ m s^{-1}. The two particles collide and coalesce.

After the collision they move with velocity $\begin{bmatrix} 2 \\ 8 \end{bmatrix}$ m s^{-1}.

  (a) Find *m*.
  (b) Find *k*.

Test yourself ANSWERS

5 (a) $m = 5$ kg;   **(b)** $k = 2$ m s^{-1}.

4 (a) $\begin{bmatrix} 6.4 \\ 0.4 \end{bmatrix}$;   **(b)** $\begin{bmatrix} 0.25 \\ 10 \end{bmatrix}$.

3 1 m s^{-1} in same direction as before collision.

2 16 m s^{-1}.

1 12 m s^{-1}.

Exam style practice papers

MM1A

Time allowed 1 hour 15 minutes

Answer **all** questions

---

1 A ball is dropped from a height of 8 metres. The ball falls vertically until it hits the ground.

    (a) Find the speed of the ball when it hits the ground.   (2 marks)

    (b) Find the time that it takes the ball to reach the ground.   (2 marks)

    (c) State the key assumption that you have made about the motion of the ball.   (1 mark)

    (d) Sketch a velocity–time graph for the motion of the ball.   (2 marks)

2 Two particles, A and B, are moving on a horizontal surface when they collide. Before the collision the velocity of A is $\begin{bmatrix} 3 \\ 6 \end{bmatrix}$ m s^{-1} and the velocity of B is $\begin{bmatrix} -5 \\ 4 \end{bmatrix}$ m s^{-1}. After the collision, both particles move together with velocity **v**.

    (a) If the mass of A is 2 kg and the mass of B is 3 kg, find **v**.   (3 marks)

    (b) If the particles have the same mass, find **v**.   (3 marks)

3 A box of mass 20 kg is at rest on a rough horizontal surface. Two ropes are attached to the box and exert forces as shown in the diagram.

Model the box as a particle.

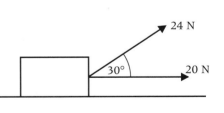

    (a) Draw a diagram to show all of the forces acting on the box.   (1 mark)

    (b) Show that the magnitude of the normal reaction force acting on the box is 184 N.   (3 marks)

    (c) Calculate the magnitude of the friction force acting on the box.   (2 marks)

    (d) The coefficient of friction between the box and the surface is μ. Find an inequality that μ must satisfy.   (3 marks)

4 A boat is to cross a river. In still water the boat would travel to 3 m s^{-1}. A current flows at a speed of 1 m s^{-1} in the river.

   (a) If the velocity of the boat relative to the water is perpendicular to the bank, find the angle between the resultant velocity of the boat and the bank.   (4 marks)

   (b) If the resultant velocity of the boat is perpendicular to the bank, find the angle between the bank and the velocity of the boat.   (4 marks)

5 The diagram shows two particles that are connected by a light, inelastic string that passes over a smooth, light pulley. One of the particles is on a rough horizontal surface and the other is hanging freely.

The particle on the surface has mass 9 kg and the other particle has mass 5 kg. The coefficient of friction between the particle and the surface is 0.2. The particles are released from rest.

   (a) Find the acceleration of the particles.   (5 marks)

   (b) Find the tension in the string.   (2 marks)

   (c) When the particles have been moving for 1 second the string breaks and the particle on the surface slows down and stops before it reaches the pulley. Find the total distance travelled by this particle.   (5 marks)

6 A golfer hits a ball from a raised platform. The ball initially moves at 45 m s^{-1} and at an angle of $50°$ above the horizontal. The ball lands on a horizontal surface that is 5 m lower than the platform. The diagram shows the path of the ball.

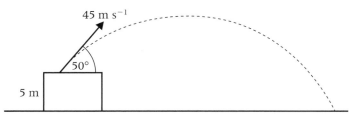

   (a) Show that the time that the ball is in the air is 7.18 seconds, correct to three significant figures.   (5 marks)

   (b) Find the horizontal distance travelled by the ball.   (2 marks)

7 A speed boat moves with a constant acceleration of $(-\mathbf{i}+\mathbf{j})\,\text{m s}^{-2}$. Initially the boat is at the origin and has velocity $4\mathbf{i}\,\text{m s}^{-1}$. The unit vectors $\mathbf{i}$ and $\mathbf{j}$ are directed east and north, respectively. The velocity of the boat at time t seconds is $\mathbf{v}\,\text{m s}^{-1}$.

   (a) Find $\mathbf{v}$ in terms of t.   (2 marks)

   (b) At time T seconds, where $T > 0$, the speed of the boat is $20\,\text{m s}^{-1}$.

      (i) Write down an equation that T must satisfy.   (3 marks)

      (ii) Hence, find T.   (2 marks)

      (iii) Find the distance of the boat from the origin at this time.   (4 marks)

MMIB

Time allowed 1 hour 30 minutes

Answer **all** questions

---

1 A car, of mass 1200 kg, accelerates from rest to a speed of $10\,\text{m s}^{-1}$, as it travels a distance of 40 m. Assume that the acceleration of the car is constant.

   (a) Calculate the acceleration of the car and the magnitude of the resultant force on the car.   (4 marks)

   (b) If the car continues with the same acceleration, how long would it take to reach a speed of $20\,\text{m s}^{-1}$ and how far would it have travelled since setting off.   (4 marks)

   (c) Explain why it is unlikely that the car would continue to move with the same acceleration.   (2 marks)

2 Two particles, A and B, have masses of 3 kg and 7 kg, respectively. They are moving on a horizontal surface when they collide. Just before the collision the velocity of A is $\begin{bmatrix} 2 \\ -6 \end{bmatrix}\,\text{m s}^{-1}$ and the velocity of B is $\begin{bmatrix} -3 \\ 4 \end{bmatrix}\,\text{ms}^{-1}$.

   (a) Find the velocity of the particles after the collision if they coalesce.   (3 marks)

   (b) Find the velocity of A after the collision, if the velocity of B is $\begin{bmatrix} 0 \\ 1 \end{bmatrix}\,\text{m s}^{-1}$.   (3 marks)

3 Two particles of mass 3 kg and 7 kg are joined by a light, inextensible string that passes over a light, smooth pulley. Assume that there is no air resistance.

The particles are initially released from rest at the same level.

   (a) By forming an equation of motion for each particle, find the acceleration of the particles.   (4 marks)

   (b) Find the tension in the string.   (2 marks)

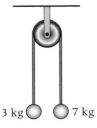

3 kg   7 kg

(c) Find the speed of the particles when they are 20 cm apart. (2 marks)

(d) If the string was not light, what would happen to the acceleration of the particles? (2 marks)

4 A car, of mass 1000 kg, skids for 20 m and then hits a car of mass 1200 kg that is moving at $10 \, \text{m s}^{-1}$ in the same direction. After the collision, both cars move together at a speed of $12 \, \text{m s}^{-1}$. The coefficient of friction between the tyres and the road for the first car is 0.8. Assume that all the motion takes place on a horizontal surface.

(a) Use conservation of momentum to find the speed of the first car, just before the collision. (2 marks)

(b) Find the speed of the first car when it begins to skid. (4 marks)

5 An aeroplane heads due north at a speed of $200 \, \text{m s}^{-1}$. A wind is blowing north east and has speed $50 \, \text{m s}^{-1}$.

(a) Sketch a velocity triangle. (1 mark)

(b) Calculate the resultant speed of the aeroplane. (4 marks)

(c) Find the bearing on which the aeroplane actually moves. Give your answer to the nearest $0.1°$. (4 marks)

6 A child pulls a sledge, of mass 20 kg, along a snow-covered surface. The child uses a rope that remains horizontal as he pulls. The coefficient of friction between the ground and the sledge is 0.2. The tension in the rope is T N.

(a) The sledge is pulled along a horizontal surface.

   (i) Show that the magnitude of the friction force acting on the sledge is 39.2 N. (2 marks)

   (ii) Find the tension in the rope if the sledge accelerates at $0.05 \, \text{m s}^{-2}$ on a horizontal surface. (4 marks)

   (iii) State the tension needed to keep the sledge moving at a constant speed. (2 marks)

(b) The child then pulls the sledge up a slope inclined at an angle of $10°$ to the horizontal. The rope is parallel to the slope. Find the tension if the sledge moves at a constant speed. (4 marks)

7 A particle is initially at rest and has position vector $(30\mathbf{i} + 400\mathbf{j})$ m, where the unit vectors $\mathbf{i}$ and $\mathbf{j}$ are east and north, respectively. The particle moves with a constant acceleration so that 10 seconds later its position is $(100\mathbf{i} + 450\mathbf{j})$ m.

(a) (i) Show that the acceleration of the particle is $(1.4\mathbf{i} + \mathbf{j}) \, \text{m s}^{-2}$. (3 marks)

   (ii) Find the velocity of the particle after 10 s. (2 marks)

   (iii) Find an expression for the position of the particle at time t s. (2 marks)

(b) After the first 10 seconds the particle stops accelerating and moves with a constant velocity. Find the position vector of the particle when it has been moving for 25 s.

(4 marks)

8 A shot is thrown with an initial velocity $10 \, \mathrm{m \, s^{-1}}$, at an angle of 40° above the horizontal.

(a) A simple model assumes that the height of release of the shot is zero. Calculate the range of the shot. (5 marks)

(b) In fact the shot is released at a height of 1.8 m. What would happen if the shot was thrown in this way inside a hall of length 20 m and height 3.5 m? (6 marks)

Answers

1 Mathematical modelling in mechanics

1 Increase distance by between 2 to 5 metres depending on size of cars.

2 This would reduce the distance between the humps.

3 No air resistance, no lift as no motion due to spin, particle, ignore initial height.

4 (a) For example:
        No air resistance
        Light rod

   (b) For example:
        Length of rod
        Acceleration due to gravity
        Mass of pendulum
        Speed of pendulum
        Angle between pendulum and the vertical

5 (a) lower mass falls;

   (b) no change.

6 Parachutist hits the ground at a very high speed.

2 Kinematics in one dimension

1 $84\,\text{m}$, $1.6\,\text{m s}^{-2}$, $0\,\text{m s}^{-2}$, $-2\,\text{m s}^{-2}$.

2 $104\,\text{m}$.

3 (a) Motion with constant positive velocity;

   (b) Motion with constant positive velocity, (perfectly elastic collision occurs?), then motion in opposite direction with the same speed but with the direction reversed;

   (c) The body starts with positive velocity which increases until it reaches a maximum. The velocity then decreases until the body is at rest instantaneously at the highest point on the graph before returning in the direction from which it came with increasing speed;

   (d) The body starts with positive velocity which decreases until it is at an instantaneous rest at the highest point on the graph. The body then speeds up in the opposite direction until it reaches a maximum velocity before slowing to an instantaneous rest at the lowest point of the curve. It then returns in its original direction, speeding up to a maximum velocity before again decelerating to an instantaneous rest at the next highest point of the graph. The motion continues cyclically.

4 **(a)**

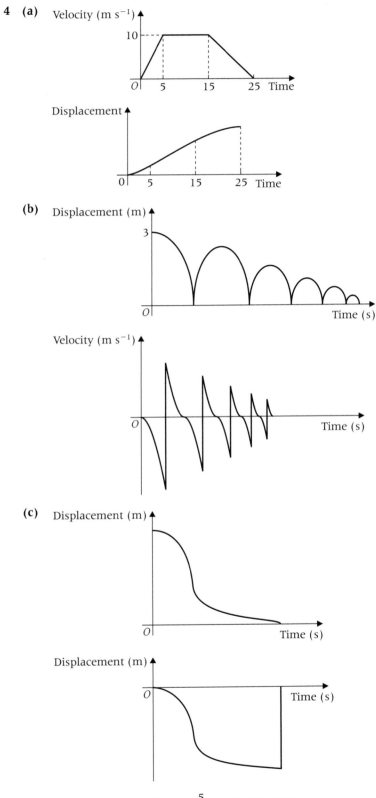

5 **(a)** $0 \, \text{m s}^{-2}$; **(b)** $2 \, \text{m s}^{-2}$; **(c)** $-\dfrac{5}{3} \, \text{m s}^{-2}$; **(d)** $437.5 \, \text{m}$.

6 (a)

(b) 584 m.

7 0.25 m s^{-2}, 0 m s^{-2}, −0.5 m s^{-2}, 8000 m.

8 (a) (i) It has a constant gradient (i.e. it is a straight line), **(ii)** $\frac{2}{3}$ m s^{-2};
(b) $T = 21$ s.

9 (a) 80 m; **(b)** $2\frac{2}{3}$ m s^{-1}; **(c)** $1\frac{1}{3}$ m s^{-1}.

10 (a) 0.3 m s^{-1}; **(b)** −0.1 m s^{-1}.

11 (a) $t = 4$ s, $t = 5.6$ s; **(b)** 50 m; **(c)** 58 m.

12 (a)

(b) 10 m s^{-1}, 2.5 m s^{-2}; **(c)** $\frac{16}{9}$ s, 5.06 m s^{-2}

13 (a) (i) $\frac{2U}{3}$ m s^{-2}, **(ii)** $12U$ m; **(b)** 2 m s^{-1}.

EXERCISE 2B

1 (a) 100 m, 20 m s^{-1}; **(b)** 22.5 m s^{-1}, 206.25 m.

2 (a) 0.3 m s^{-2}; **(b)** 33.3 s.

3 (a) 0.7 m s^{-2}; **(b)** 85 m.

4 (a) 0.2 m s^{-2}; **(b)** 10 m; **(c)** 15 m.

5 (a) 0.417 m s^{-2}; **(b)** 361 m.

6 −10.5 m s^{-2}, 1.69 s.

7 (a) 889 m; **(b)** 66.7 s.

8 1200 m.

9 (a) 2.92 m s^{-2}, 18.7 m s^{-1}; **(b)** 8.55 s.

10 (a) 16.3 m s^{-1}; **(b)** 2.21 m s^{-2}.

11 0.580 m s^{-2}, 15.3 m s^{-1}.

12 (a) 16 m; **(b)** 8 m s^{-1}; **(c)** 14.5 s.

13 82.5 m; **(b)** 35 s; **(c)** 2.36 m s^{-1}.

14 10.5 s.

15 12 400 m.

16 (a) 24 m s^{-1}; **(b)** 7800 m.

17 8 m, 7.5 s.

18 **(a)** 25 m; **(b)** 20 s; **(c)** 50 m s⁻¹, 62.5 m s⁻¹.

19 **(a)**

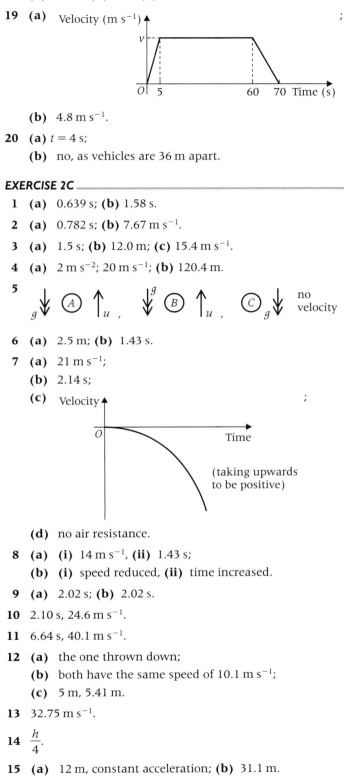

(b) 4.8 m s⁻¹.

20 **(a)** $t = 4$ s;

   (b) no, as vehicles are 36 m apart.

EXERCISE 2C

1 **(a)** 0.639 s; **(b)** 1.58 s.

2 **(a)** 0.782 s; **(b)** 7.67 m s⁻¹.

3 **(a)** 1.5 s; **(b)** 12.0 m; **(c)** 15.4 m s⁻¹.

4 **(a)** 2 m s⁻²; 20 m s⁻¹; **(b)** 120.4 m.

5

6 **(a)** 2.5 m; **(b)** 1.43 s.

7 **(a)** 21 m s⁻¹;

   (b) 2.14 s;

   (c)

   (d) no air resistance.

8 **(a)** **(i)** 14 m s⁻¹, **(ii)** 1.43 s;

   (b) **(i)** speed reduced, **(ii)** time increased.

9 **(a)** 2.02 s; **(b)** 2.02 s.

10 2.10 s, 24.6 m s⁻¹.

11 6.64 s, 40.1 m s⁻¹.

12 **(a)** the one thrown down;

   (b) both have the same speed of 10.1 m s⁻¹;

   (c) 5 m, 5.41 m.

13 32.75 m s⁻¹.

14 $\frac{h}{4}$.

15 **(a)** 12 m, constant acceleration; **(b)** 31.1 m.

16 **(a)** 0.569 m; **(b)** over-estimate due to air resistance.

3 Kinetics in two dimensions

EXERCISE 3A

1 (a) $0i + 0j$, $30i + 20.1j$, $60i + 30.4j$, $90i + 30.9j$, $120i + 21.6j$, $150i + 2.5j$, $180i - 26.4j$;

(b)

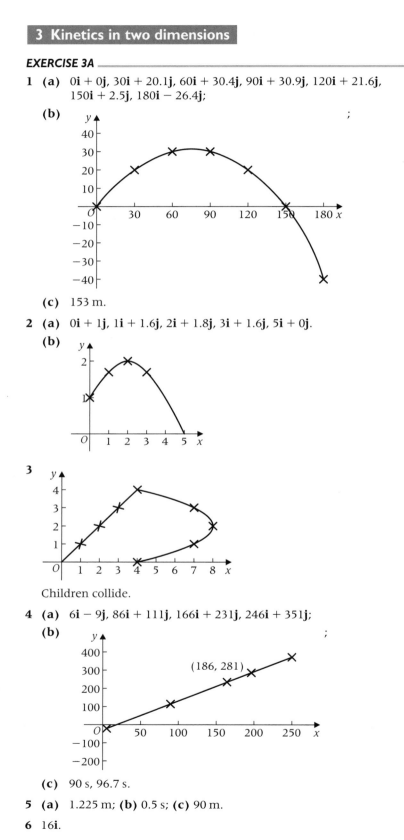

(c) 153 m.

2 (a) $0i + 1j$, $1i + 1.6j$, $2i + 1.8j$, $3i + 1.6j$, $5i + 0j$.

(b)

3

Children collide.

4 (a) $6i - 9j$, $86i + 111j$, $166i + 231j$, $246i + 351j$;

(b)

(186, 281)

(c) 90 s, 96.7 s.

5 (a) 1.225 m; (b) 0.5 s; (c) 90 m.

6 $16i$.

EXERCISE 3B

1 (a) $45\cos10°\mathbf{i} + 45\sin10°\mathbf{j} = 44.3\mathbf{i} + 7.81\mathbf{j}, \begin{bmatrix} 44.3 \\ 7.81 \end{bmatrix}$;

   (b) $105\cos30°\mathbf{i} + 105\sin30°\mathbf{j} = 90.9\mathbf{i} + 52.5\mathbf{j}, \begin{bmatrix} 90.9 \\ 52.5 \end{bmatrix}$;

   (c) $-21\cos70°\mathbf{i} + 21\sin70°\mathbf{j} = -7.18\mathbf{i} + 19.7\mathbf{j}, \begin{bmatrix} -7.18 \\ 19.7 \end{bmatrix}$;

   (d) $-62\cos10°\mathbf{i} - 62\sin10°\mathbf{j} = -61.1\mathbf{i} - 10.8\mathbf{j}, \begin{bmatrix} -61.1 \\ -10.8 \end{bmatrix}$;

   (e) $290\cos72°\mathbf{i} - 290\sin72°\mathbf{j} = 89.6\mathbf{i} - 275.8\mathbf{j}, \begin{bmatrix} 89.6 \\ -275.8 \end{bmatrix}$.

2 $-6\cos45°\mathbf{i} - 6\sin45°\mathbf{j} = -4.24\mathbf{i} - 4.24\mathbf{j}$.

3 (a) $5\mathbf{i}$; (b) $-5\mathbf{j}$; (c) $-5\mathbf{i}$; (d) $5\cos45°\mathbf{i} - 5\sin45°\mathbf{j} = 3.54\mathbf{i} - 3.54\mathbf{j}$;
   (e) $-5\cos45°\mathbf{i} + 5\sin45°\mathbf{j} = -3.54\mathbf{i} + 3.54\mathbf{j}$.

4 $8\cos50°\mathbf{i} + 8\sin50°\mathbf{j} = 5.14\mathbf{i} + 6.13\mathbf{j}$.

5 (a) 8.06 m s^{-1}, 029.7°; (b) 7.81 m s^{-1}, 140.2°; (c) 12.0 m s^{-1}, 221.6°;
   (d) 14.4 m s^{-1}, 303.7°.

6 (a) 9.85 m s^{-1}, 024.0°; (b) 3.61 m s^{-1}, 146.3°; (c) 4.47 m s^{-1}, 296.6°;
   (d) 10.6 m s^{-1}, 228.8°.

7 (a) $5\mathbf{i} - 6\mathbf{j}$; (b) $-\mathbf{i} + 5\mathbf{j}$; (c) $\mathbf{i} + 10\mathbf{j}$; (d) $7\mathbf{i} - \mathbf{j}$; (e) $13\mathbf{i} - 10\mathbf{j}$;
   (f) $32\mathbf{i} - 50\mathbf{j}$.

8 (a) $\begin{bmatrix} 4 \\ -1 \end{bmatrix}$; (b) $\begin{bmatrix} -2 \\ -3 \end{bmatrix}$; (c) $\begin{bmatrix} -5 \\ 5 \end{bmatrix}$; (d) $\begin{bmatrix} -6 \\ 11 \end{bmatrix}$; (e) $\begin{bmatrix} 2 \\ -18 \end{bmatrix}$; (f) $\begin{bmatrix} 2 \\ 17 \end{bmatrix}$.

9 (a) 8.54 m; (b) 159.4°.

10 3.61 m s^{-2}, 303.7°.

EXERCISE 3C

1 $-0.3\mathbf{i} - 0.5\mathbf{j}$.

2 (a) $4.8\mathbf{i} + 2.4\mathbf{j}$; (b) $31.7\mathbf{i} + 8.6\mathbf{j}$.

3 $\mathbf{v} = (3 + t)\mathbf{i} + (-5 + t)\mathbf{j}$, $\mathbf{r} = \left(3t + \dfrac{t^2}{2}\right)\mathbf{i} + \left(-5t + \dfrac{t^2}{2}\right)\mathbf{j}$.

4 $36\mathbf{i} + 48\mathbf{j}$, $108\mathbf{i} + 144\mathbf{j}$.

5 $\mathbf{r} = 4t\mathbf{i} + (2 + 9t - 5t^2)\mathbf{j}$, $8\mathbf{i}$.

6 $\mathbf{r} = (0.2 + \sqrt{3}t)\mathbf{i} + (0.1 + t)\mathbf{j}$, $1.76\mathbf{i} + 1\mathbf{j}$.

7 (b) 6.59 s, $\mathbf{r} = 46.1\mathbf{i}$; (c) 33.3 m s^{-1}.

8 (a) 6 s, $120\mathbf{i}$, 36.1 m s^{-1}; (b) 4 s.

9 (a) $\mathbf{v} = (4 - 0.02t)\mathbf{i} + (6 - 0.04t)\mathbf{j}$, $\mathbf{r} = (80 + 4t - 0.01t^2)\mathbf{i} +$
   $(20 + 6t - 0.02t^2)\mathbf{j}$; (b) $416\mathbf{i} + 92\mathbf{j}$; (c) $0.4\mathbf{i} - 1.2\mathbf{j}$.

10 (a) $\mathbf{r}_H = 6t\mathbf{i} + (80 - 3t)\mathbf{j}$, $\mathbf{r}_A = (5t + 0.05t^2)\mathbf{i} + (10 + 0.025t^2)\mathbf{j}$;
   (b) 20 s, $120\mathbf{i} + 20\mathbf{j}$.

11 $\mathbf{v} = (2 - 0.06t)\mathbf{i} + (3 - 0.04t)\mathbf{j}$, $\mathbf{r} = (40 + 2t - 0.03t^2)\mathbf{i} +$
   $(20 + 3t - 0.02t^2)\mathbf{j}$; (b) 60 s.

12 (a) 68.8 m; (b) (i) $t = 200 \text{ s}$, (ii) $t = 66\frac{2}{3} \text{ s}$.

13 **(a)** $1.2\mathbf{i} + 0.8\mathbf{j}$; **(b)** $405\mathbf{i} + 90\mathbf{j}$.

14 **(a)** $4\mathbf{j}$; **(b)** $0.721\,\text{m s}^{-2}$; **(c)** $30\mathbf{i} + 20\mathbf{j}$.

15 **(a)** $3\mathbf{i}$; **(b)** $\mathbf{v} = (3 + 0.1t)\mathbf{i} + 0.2t\mathbf{j}$; **(c)** $3 + 0.1T = 0.2T$; **(d)** $89.4\,\text{m}$.

16 **(a)** $a = 2$; **(b)** **(i)** $4\mathbf{i} + 12\mathbf{j}$, **(ii)** $-2\mathbf{i} - 4\mathbf{j}$; **(c)** $16.5\,\text{m}$.

17 **(a)** $40\mathbf{i} - 5\mathbf{j}$; **(b)** $3\mathbf{i} - \mathbf{j}$; **(c)** $\mathbf{i} - 2\mathbf{j}$; **(d)** $10\mathbf{i} - 20\mathbf{j}$.

EXERCISE 3D

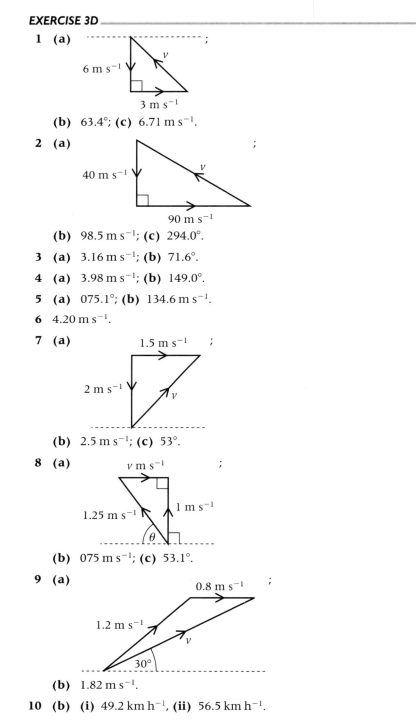

1 **(a)**

(b) $63.4°$; **(c)** $6.71\,\text{m s}^{-1}$.

2 **(a)**

(b) $98.5\,\text{m s}^{-1}$; **(c)** $294.0°$.

3 **(a)** $3.16\,\text{m s}^{-1}$; **(b)** $71.6°$.

4 **(a)** $3.98\,\text{m s}^{-1}$; **(b)** $149.0°$.

5 **(a)** $075.1°$; **(b)** $134.6\,\text{m s}^{-1}$.

6 $4.20\,\text{m s}^{-1}$.

7 **(a)**

(b) $2.5\,\text{m s}^{-1}$; **(c)** $53°$.

8 **(a)**

(b) $075\,\text{m s}^{-1}$; **(c)** $53.1°$.

9 **(a)**

(b) $1.82\,\text{m s}^{-1}$.

10 **(b)** **(i)** $49.2\,\text{km h}^{-1}$, **(ii)** $56.5\,\text{km h}^{-1}$.

4 Forces

EXERCISE 4A

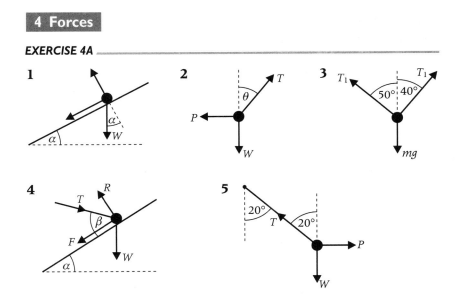

EXERCISE 4B

1 **(a)** 5 N, 36.9°; **(b)** 14 N, 38.2°; **(c)** 12.8 N, 43.0°; **(d)** 4.35 N, 151.4°;
   (e) 3.78 N, 95.1°.

2 **(a)** 90°; **(b)** 91.5°; **(c)** 152.3°; **(d)** 158.2°.

3 11 N, 1 N.

4 15.8 N.

5 11.1 N.

6 158.2°.

7 7.44 N, 53.8°.

EXERCISE 4C

1 **(a)** 1 N; **(b)** 5.96 N.

2 4.24 N.

3 $a = -7, b = -5$.

4 **(a)** $9\mathbf{i} + 17\mathbf{j}$; **(b)** 19.2 N; **(c)** 62.1°.

5 3.61 N, 123.7°.

6 4.12 N.

7 $a = 2$ or 0.4.

8 $(\lambda - 4)\mathbf{i} + 4\mathbf{j}$, $\sqrt{\lambda^2 - 8\lambda + 32}$, $\lambda = 1$ or 7, 53.1°.

EXERCISE 4D

1 **(a)** 4.50, 5.36; **(b)** -7.25, 3.38; **(c)** -5.79, -6.89.

2 **(a)** $4.50\mathbf{i} + 5.36\mathbf{j}$; **(b)** $-7.25\mathbf{i} + 3.38\mathbf{j}$; **(c)** $-5.79\mathbf{i} - 6.89\mathbf{j}$.

3 **(a)** $\begin{bmatrix} 4.50 \\ 5.36 \end{bmatrix}$; **(b)** $\begin{bmatrix} -7.25 \\ 3.38 \end{bmatrix}$; **(c)** $\begin{bmatrix} -5.79 \\ -6.89 \end{bmatrix}$.

4 **(a)** $-0.5Mg$, $-0.866Mg$; **(b)** $-W\sin\alpha$, $-W\cos\alpha$.

5 **(a)** 5.92 N, 88.5° below *x*; **(b)** 1.43 N, 82.3° below *x*;
   (c) 7.21 N, 46.3° below *x*; **(d)** 2.49 N, 139.2° above *x*.

6 0 N.

7 **(a)** $-0.707W, -0.707W$; **(b)** $-0.5W, -0.866W$; **(c)** $-0.866W, 0.5W$.

8 $mg\sin\alpha$.

EXERCISE 4E

1 **(a)** $F_1 = 1.71$ N, $F_2 = 4.70$ N; **(b)** $F_1 = 4.01$ N, $F_2 = 17.7$ N;
   (c) $F_1 = 5.22$ N, $F_2 = 7.04$ N.

2 **(a)** $F = 25$ N, $\theta = 73.7°$; **(b)** $F = 9.42$ N, $\theta = 15.8°$;
   (c) $F = 8.64$ N, $\theta = 130.4°$.

3 **(a)** $F_1 = 7.83$ N, $F_2 = 10.21$ N; **(b)** $F_1 = 6.77$ N, $F_2 = 9.44$ N;
   (c) $F_1 = 6.72$ N, $F_2 = 16.4$ N; **(d)** $F_1 = 10$ N, $F_2 = 8.34$ N;
   (e) $F_1 = 4.79$ N, $F_2 = 6.16$ N; **(f)** $F_1 = 131$ N, $F_2 = 176$ N.

4 **(a)** $-8\mathbf{i} - 4\mathbf{j}$; **(b)** 8.94 N, 153.4° below **i**.

5 $a = -11, b = 0$.

6 **(a)** $\mathbf{F}_2 = -8\mathbf{i} - 9\mathbf{j}$; **(b)** $\mathbf{F}_3 = 2\mathbf{i} - 2\mathbf{j}$.

7 $x = 1, y = -1.5$.

8 2.24.

9 Yes.

10 24.5 N, 42.4 N.

11 38.7 N, 72.7 N.

12 5.57 N, 17.9°.

13 $P = 8.39$ N, $R = 13.1$ N.

14 11.2 N, 26.6°.

15 124.2°, 97.2°, 138.6°.

16 **(b)** 283 N.

17 **(a)** 61.8 N; **(b)** 8.16 kg.

18 27.1 N, 24.0 N.

19 **(b)** 3.46 N.

20 **(a)** $\begin{bmatrix} 21.7 \\ 12.5 \end{bmatrix}$; **(b)** $\begin{bmatrix} -7.51 \\ -26.6 \end{bmatrix}$.

21 $-88\mathbf{i} - 128\mathbf{j}$, 155 N.

EXERCISE 4F

1 0.0255

2 **(a)** no motion; **(b)** slides; **(c)** no motion.

3 0.567.

4 73.6 N.

5 51.8 N, 51.8 N, 0.530.

6 88.2 N.

7 (a) 0.0850;

(b) 30.8 N.

9 (a)

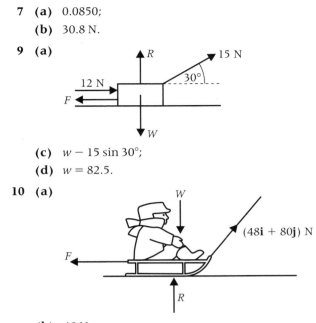

(c) $w - 15 \sin 30°$;

(d) $w = 82.5$.

10 (a)

(b) 48 N;

(d) 200 N.

EXERCISE 4G

1 (a) slides; **(b)** no motion; **(c)** slides.

2 (a) 7.37 N; **(b)** 25.0 N; **(c)** 41.9 N.

3 (a) 26.7 N; **(b)** 47.2 N; **(c)** 59.4 N.

4 34.6 N.

5 (a) 28.3 N; **(b)** 181 N; **(c)** 671 N.

6 (a) 29.2 N; **(b)** 7.88 N; **(c)** 14.5 N.

7 $F = 21.9$ N, $R = 77.9$ N, 0.281.

8 (a)

$R = 53.9$ N, $F = 33.6$ N, $\mu \geq 0.623$;

(b)

$R = 36.8$ N, $F = 6.59$ N, $\mu \geq 0.179$;

(c)

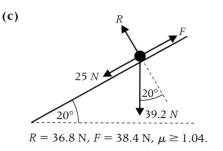

$R = 36.8\,\text{N}$, $F = 38.4\,\text{N}$, $\mu \geq 1.04$.

5 Newton's laws of motion

EXERCISE 5A

1 **(a)** $X = 6\,\text{N}$, $Y = 5\,\text{N}$; **(b)** $X = 10\,\text{N}$, $Y = 12\,\text{N}$; **(c)** $X = 69.3\,\text{N}$, $Y = 69.3\,\text{N}$.

2 **(a)** $10\,\text{N}$, $36.9°$; **(b)** $15.7\,\text{N}$, $60°$; **(c)** $8.66\,\text{N}$, $60°$.

3 $1350\,\text{N}$.

4 0.245.

5 **(a)** $T = 115\,\text{N}$; **(b)** $R = 112\,\text{N}$.

6 $187\,\text{N}$.

7 $29.9\,\text{m s}^{-1}$.

8 $17.9\,\text{m s}^{-1}$.

9 $53.9\,\text{m s}^{-1}$.

EXERCISE 5B

1 $0.5\,\text{m s}^{-2}$.

2 $6\,\text{N}$.

3 $(2.5\mathbf{i} + \mathbf{j})\,\text{m s}^{-2}$.

4 $0.7\,\text{m s}^{-2}$.

5 **(a)** $2970\,\text{N}$; **(b)** $2880\,\text{N}$; **(c)** $2910\,\text{N}$.

6 **(a)** $a = 8\,\text{m s}^{-2}$; **(b)** $a = 0.8\,\text{m s}^{-2}$, $R = 98\,\text{N}$; **(c)** $T = 47.3\,\text{N}$, $R = 74.3\,\text{N}$; **(d)** $103\,\text{N}$; **(e)** $\theta = 30°$, $a = 0.329\,\text{m s}^{-2}$; **(f)** $T = 66.5\,\text{N}$, $a = 3\,\text{m s}^{-2}$.

7 **(a)**

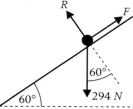

(b) $147\,\text{N}$; **(d)** $5.55\,\text{m s}^{-2}$; **(e)** child is a particle.

8 **(a)** **(ii)** $2.78\,\text{m s}^{-2}$; **(b)** **(i)** $69.5\,\text{N}$, **(ii)** $5.27\,\text{m s}^{-1}$.

9 $4167\,\text{N}$.

10 $4.9\,\text{m s}^{-2}$.

11 $4\,\text{s}$.

12 $33.3\,\text{N}$, 0.195.

13 $11.8°$.

14 (a) (i) 74.7 N, **(ii)** 90.7 N;
  (b) (i) 20 s;
    (ii)

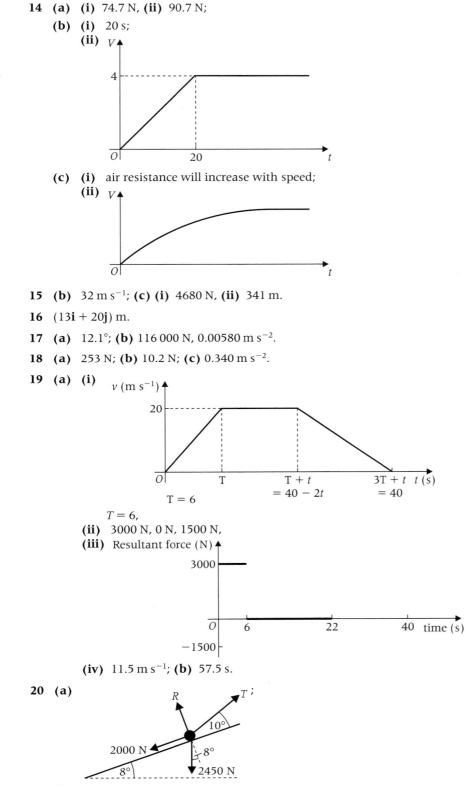

  (c) (i) air resistance will increase with speed;
    (ii)

15 (b) 32 m s^{-1}; **(c) (i)** 4680 N, **(ii)** 341 m.

16 $(13\mathbf{i} + 20\mathbf{j})$ m.

17 (a) 12.1°; **(b)** 116 000 N, 0.00580 m s^{-2}.

18 (a) 253 N; **(b)** 10.2 N; **(c)** 0.340 m s^{-2}.

19 (a) (i)

$T = 6$,
  (ii) 3000 N, 0 N, 1500 N,
  (iii) Resultant force (N)

  (iv) 11.5 m s^{-1}; **(b)** 57.5 s.

20 (a)

  (b) 2428 N;
  (c) The resistance force will increase and it will stop accelerating.

21 (a) (i) $6\,\text{m s}^{-1}$, **(ii)** $-12\,\text{m s}^{-2}$;
   (c) $7.6\,\text{m s}^{-2}$;
   (d) $1.13\,\text{s}$.

22 (b) 4540 N, 4280 N;
   (c) The 4540 N would increase and the 4280 N force would decrease.

23 (a) There is a currently a resultant force of $441 \sin 40° - 89 = 194\,\text{N}$ (3 s.f.) down the slide. This will produce an acceleration, therefore there must be air resistance of 194 N to counter this;
   (b) $3.24\,\text{m s}^{-2}$.

24 (a) (i)

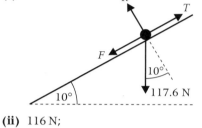

   (ii) 116 N;
   (b) $1.20\,\text{m s}^{-2}$;
   (c) sledge is a particle, no air resistance.

EXERCISE 5C

1 The hard object exerts an equal and opposite force back on your hand!

2 (a)

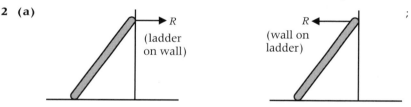

   (b) If they are not equal and opposite, then there is a resultant force at the point and the ladder will move. Newton's third law says that they exert equal and opposite forces on each other.

3 The higher the table jumped from, the larger the forces involved. Before landing there is no force between the child and the ground. On impact, the force exerted on the child by the ground increases and is upwards. Newton's third law says the force on the ground by the child is equal in size but is downwards. The magnitude of these forces increases (to decelerate the child) and then decreases again until the force exerted upwards by the ground on the child balances the child's weight.

4

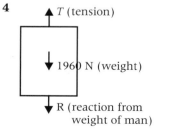

$T = 2863\,\text{N}$, $R = 803\,\text{N}$, $W = 1960\,\text{N}$.

5 For the situation on a straight track shown

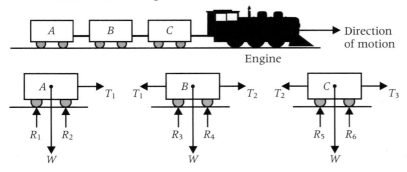

The weights are the same provided the carriages are identical, then also R1 = R3 = R5 and R2 = R4 = R6. These will all be equal if the centre of mass is symmetrical between the wheels. All the tensions in the couplings are the same when the train travels at constant velocity.

6 Connected particles

EXERCISE 6A

1. 2.45 m s^{-2}, 36.75 N.

2. **(a)** 1.09 m s^{-2}, 43.6 N; **(b)** 1.96 m s^{-2}, 47.0 N; **(c)** 3.77 m s^{-2}, 54.3 N.

3. **(a)** 2.67 m s^{-2}; **(b)** 0.612 s.

4. $\dfrac{g}{5} \text{ m s}^{-2}$, $\dfrac{6}{5}mg \text{ N}$.

5. 0.25.

6. 0.98 m s^{-2}, 17.6 N.

7. **(a)** 2.43 m s^{-2}; **(b)** 22.1 N; **(c)** -2 m s^{-2}.

8. **(a)** 1100 N; **(b)** 400 N.

9. 1.93 m s^{-2}, 15.7 N, 2.78 m s^{-1}.

10. **(b)** 47.0 N.

11. **(a)** **(ii)** 3.36 N; **(b)** 2.8 m.

12. **(a)** 29.4 N; **(c)** 35.3 N.

13. **(b)** 0.8 m s^{-2}; **(c)** 1.8 N; **(d)** 1.25 s.

14. **(b)** 38 N; **(c)** 36.2 N; **(d)** 0.62.

15. **(a)** **(i)** mg, **(ii)** $\mu \geqslant \dfrac{1}{2}$; **(b)** $\dfrac{2mg}{3}$.

16. **(a)** **(ii)** 336 N; **(b)** smooth pulley, light string; **(c)** 2.65 m s^{-1}.

17. **(a)** **(ii)** 470 N; **(b)** 2.4 m.

18. **(a)** **(i)** 25 N, **(ii)** 150 N; **(b)** 100 N.

19. **(b)** $\dfrac{mMg(1 + \mu)}{m + M}$; **(c)** $\sqrt{\dfrac{m + M}{g(m - \mu M)}}$, $\sqrt{\dfrac{g(m - \mu M)}{m + M}}$.

20. **(a)** 3.92 m s^{-2}; **(b)** 27.6 N.

21. **(a)** 0.668 m s^{-2}; **(b)** 9.48 s.

22. **(a)** 0.942 m s^{-2}; **(b)** 0.250 m s^{-2}.

23. **(b)** 31.3 N.

7 Projectiles

EXERCISE 7A

1 (a) 2.60 s; (b) 33.1 m; (c) 8.27 m.

2 19.9 m.

3 3.18 m.

4 15.3 m, 17.7 m.

5 Yes, predict 108 m, unlikely to cause a problem.

6 (a) 1.22 s, 1.84 m, 12.7 m;
   (b) 10.5 m s^{-1}, 6.0°, 11.1 m s^{-1}, 20.1° below the horizontal.

7 (b) 39.2 m; (c) (i) 12.0 m s^{-1}, (ii) horizontal.

8 (a) 16.9 m; (b) 39.0 m; (c) 10.5 m s^{-1}.

9 (b) (i) 3.025 m, (ii) 0.714 s, (iii) 1.57 s.

10 (a) $u \cos \alpha$, $u \sin \alpha - gt$; (b) $u \cos \alpha t$, $u \sin \alpha t - \dfrac{1}{2}gt^2$; (c) $\dfrac{u^2 \sin^2 \alpha}{2g}$.

EXERCISE 7B

1 (a) 12.0 m; (b) 4.11 m.

2 (a) 40.2 m; (b) 42.5 m.

3 (a) 100 m; (b) 4.94 s, 98.8 m.

4 3.05 s, 52.8 m.

5 (c) 82 m.

6 (b) under the bar; (c) 9.57 m s^{-1}.

7 Yes, passes over at a height of 3.02 m.

8 (a) $18\frac{1}{8}$ m; (b) 21.7 m.

9 23.6 m s^{-1}.

10 (a) 25.3 m s^{-1}; (b) clears net by 31.6 cm.

11 (a) 3.45 m; (b) 1.01 s; (c) 4.04 m.

12 (a) 2.63 m; (b) 2.77 m.

13 (a) no air resistance, horizontal ground; (b) 105 m; (c) 12.3 m.

14 (a) (ii) $\sqrt{540} \text{ m s}^{-1}$; (b) yes.

15 (a) (ii) 26 m s^{-1}.

16 (b) $\dfrac{24U^2}{g}$; (c) $5U$; (d) $\dfrac{U}{g}, \dfrac{7U}{g}$.

17 (b) 105 m; (c) 45.7 m s^{-1}.

8 Momentum

EXERCISE 8A

1 (a) 4 800 000 N s; (b) 0.012 N s; (c) 12 000 N s.

2 $3\frac{1}{3} \text{ m s}^{-1}$.

3 2.75 m s^{-1}.

4 $4\,\mathrm{m\,s^{-1}}$.

5 $2\,\mathrm{m\,s^{-1}}$.

6 $3\,\mathrm{m\,s^{-1}}$.

7 $2.14\,\mathrm{m\,s^{-1}}$.

8 $0.05\,\mathrm{m\,s^{-1}}$.

9 $2u$ in the opposite direction.

10 **(a)** $4.90\,\mathrm{m\,s^{-1}}$; **(b)** $14.7\,\mathrm{m\,s^{-1}}$.

11 **(a)** $48.9\,\mathrm{km\,h^{-1}}$ or $13.6\,\mathrm{m\,s^{-1}}$; **(b)** $2.31\,\mathrm{m\,s^{-2}}$, $415\,000\,\mathrm{N}$.

12 **(a)** $1\,\mathrm{m\,s^{-1}}$; **(b)** $8.57\,\mathrm{cm}$.

13 $6.33\,\mathrm{m\,s^{-1}}$.

14 **(a)** $t = 4\,\mathrm{s}$; **(b)** no, as $36\,\mathrm{m}$ is the distance between the cars;
(c) $7.2\,\mathrm{m\,s^{-1}}$.

15 $v = 1$.

16 $V = 0.4$.

17 **(b)** $18.6\,\mathrm{kg}$.

18 **(b)** $1.708\,\mathrm{m\,s^{-1}}$.

19 **(a)** $8\,\mathrm{m\,s^{-1}}$; **(b) (i)** $19.6\,\mathrm{N}$, **(ii)** $16.3\,\mathrm{m}$.

20 **(a) (ii)** $-6.86\,\mathrm{m\,s^{-2}}$.; **(b)** $20.7\,\mathrm{m\,s^{-1}}$.

21 **(b)** $1.5\,\mathrm{m\,s^{-1}}$;
(c) (i) $-0.5\,\mathrm{m\,s^{-2}}$, **(ii)** $2.25\,\mathrm{m}$.

22 **(a)** $28.4\,\mathrm{m\,s^{-1}}$; **(b)** $1.136\,\mathrm{m\,s^{-1}}$; **(c)** $0.110\,\mathrm{m}$.

EXERCISE 8B

1 $(3.6\mathbf{i} - 2.8\mathbf{j})\,\mathrm{m\,s^{-1}}$, $4.56\,\mathrm{m\,s^{-1}}$.

2 $\begin{bmatrix} 8.4 \\ 4 \end{bmatrix}\,\mathrm{m\,s^{-1}}$.

3 $\begin{bmatrix} -1.5 \\ 0.25 \end{bmatrix}\,\mathrm{m\,s^{-1}}$.

4 Parallel to the unit vector $\mathbf{i}$ at $7.5\,\mathrm{m\,s^{-1}}$.

5 **(a)** 5; **(b)** $(0.6\mathbf{j})\,\mathrm{m\,s^{-1}}$.

6 $\alpha = 11$, $\beta = -\dfrac{2}{3}$.

7 $\begin{bmatrix} -20 \\ 22.5 \end{bmatrix}\,\mathrm{m\,s^{-1}}$.

8 $(0.75\mathbf{i} + 0.25\mathbf{j})\,\mathrm{m\,s^{-1}}$.

9 $(4.2\mathbf{i} - 7\mathbf{j})\,\mathrm{m\,s^{-1}}$.

10 $\begin{bmatrix} 1.92 \\ 1.28 \end{bmatrix}\,\mathrm{m\,s^{-1}}$.

11 **(a)** $\begin{bmatrix} -3 \\ 4 \end{bmatrix}$; **(b)** $5\,\mathrm{m\,s^{-1}}$; **(c)** B.

12 **(b)** $\begin{bmatrix} -1 \\ 3 \end{bmatrix}$; **(c)** $\begin{bmatrix} -1 \\ 17 \end{bmatrix}$.

Answers to practice papers

MMIA

1 (a) $12.5 \, \text{m s}^{-1}$; (b) $1.28 \, \text{s}$; (c) no air resistance;
   (d)

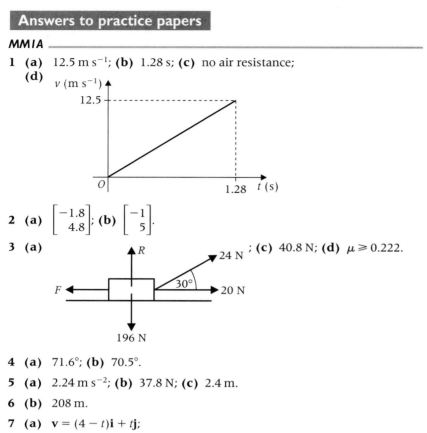

2 (a) $\begin{bmatrix} -1.8 \\ 4.8 \end{bmatrix}$; (b) $\begin{bmatrix} -1 \\ 5 \end{bmatrix}$.

3 (a) ; (c) $40.8 \, \text{N}$; (d) $\mu \geqslant 0.222$.

4 (a) $71.6°$; (b) $70.5°$.

5 (a) $2.24 \, \text{m s}^{-2}$; (b) $37.8 \, \text{N}$; (c) $2.4 \, \text{m}$.

6 (b) $208 \, \text{m}$.

7 (a) $\mathbf{v} = (4 - t)\mathbf{i} + t\mathbf{j}$;
   (b) (i) $(4 - T)^2 + T^2 = 20^2$, (ii) $T = 16$, (iii) $143 \, \text{m}$.

MMIB

1 (a) $1.25 \, \text{m s}^{-2}$, $1500 \, \text{N}$; (b) $16 \, \text{s}$, $160 \, \text{m}$;
   (c) air resistance would increase with speed and decrease the resultant force and acceleration.

2 (a) $\begin{bmatrix} -1.5 \\ 1 \end{bmatrix} \text{m s}^{-1}$; (b) $\begin{bmatrix} -5 \\ 1 \end{bmatrix} \text{m s}^{-1}$.

3 (a) $3.92 \, \text{m s}^{-2}$; (b) $41.16 \, \text{N}$; (c) $0.885 \, \text{m s}^{-1}$; (d) increases.

4 (a) $14.4 \, \text{m s}^{-1}$; (b) $22.8 \, \text{m s}^{-1}$.

5 (a) ; (b) $238 \, \text{m s}^{-1}$; (c) $008.5°$.

6 (a) (ii) $40.2 \, \text{N}$, (iii) $39.2 \, \text{N}$; (b) $72.6 \, \text{N}$.

7 (a) (ii) $14\mathbf{i} + 10\mathbf{j}$, (iii) $\mathbf{r} = (30 + 0.7t^2)\mathbf{i} + (400 + 0.5t^2)\mathbf{j}$;
   (b) $\mathbf{r} = 310\mathbf{i} + 600\mathbf{j}$.

8 (a) $10.0 \, \text{m}$; (b) hits the roof as it would rise to a height of $3.91 \, \text{m}$.

Index